Flinn Scientific
ChemTopic® Labs

The Periodic Table

Senior Editor

Irene Cesa
Flinn Scientific, Inc.
Batavia, IL

FLINN SCIENTIFIC INC.
"Your Safer Source for Science Supplies"
P.O. Box 219 • Batavia, IL 60510
1-800-452-1261 • www.flinnsci.com

ISBN 1-877991-72-4

Printed in the United States of America.

Table of Contents

Flinn ChemTopic® Labs Series Preface

Lab Manuals Organized Around Key Content Areas in Chemistry

In conversations with chemistry teachers across the country we have heard a common concern. Teachers are frustrated with their current lab manuals, with experiments that are poorly designed and don't teach core concepts, with procedures that are rigid and inflexible and don't work. Teachers want greater flexibility in their choice of lab activities. As we further listened to experienced master teachers who regularly lead workshops and training seminars, another theme emerged. Master teachers mostly rely on collections of experiments and demonstrations they have put together themselves over the years. Some activities have been passed on like cherished family recipe cards from one teacher to another. Others have been adapted from one format to another to take advantage of new trends in microscale equipment and procedures, technology innovations, and discovery-based learning theory. In all cases the experiments and demonstrations have been fine-tuned based on real classroom experience.

Flinn Scientific has developed a series of lab manuals based on these "cherished recipe cards" of master teachers with proven excellence in both teaching students and training teachers. Created under the direction of an Advisory Board of award-winning chemistry teachers, each lab manual in the Flinn ChemTopic® Labs series contains 4–6 student-tested experiments that focus on essential concepts and applications in a single content area. Each lab manual also contains 3–5 demonstrations that can be used to illustrate a chemical property, reaction, or relationship and will capture your students' attention. The experiments and demonstrations in the Flinn ChemTopic® Labs series are enjoyable, highly focused, and will give students a real sense of accomplishment.

Laboratory experiments allow students to experience chemistry by doing chemistry. Experiments have been selected to provide students with a crystal-clear understanding of chemistry concepts and encourage students to think about these concepts critically and analytically. Well-written procedures are guaranteed to work. Reproducible data tables teach students how to organize their data so it is easily analyzed. Comprehensive teacher notes include a master materials list, solution preparation guide, complete sample data, and answers to all questions. Detailed lab hints and teaching tips show you how to conduct the experiment in your lab setting and how to identify student errors and misconceptions before students are led astray.

Chemical demonstrations provide another teaching tool for seeing chemistry in action. Because they are both visual and interactive, demonstrations allow teachers to take students on a journey of observation and understanding. Demonstrations provide additional resources to develop central themes and to magnify the power of observation in the classroom. Demonstrations using discrepant events challenge student misconceptions that must be broken down before new concepts can be learned. Use demonstrations to introduce new ideas, illustrate abstract concepts that cannot be covered in lab experiments, and provide a spark of excitement that will capture student interest and attention.

Safety, flexibility, and choice

Safety always comes first. Depend on Flinn Scientific to give you upfront advice and guidance on all safety and disposal issues. Each activity begins with a description of the hazards involved and the necessary safety precautions to avoid exposure to these hazards. Additional safety, handling, and disposal information is also contained in the teacher notes.

The selection of experiments and demonstrations in each Flinn ChemTopic® Labs manual gives you the flexibility to choose activities that match the concepts your students need to learn. No single teacher will do all of the experiments and demonstrations with a single class. Some experiments and demonstrations may be more helpful with a beginning-level class, while others may be more suitable with an honors class. All of the experiments and demonstrations have been keyed to national content standards in science education.

Chemistry is an experimental science!

Whether they are practicing key measurement skills or searching for trends in the chemical properties of substances, all students will benefit from the opportunity to discover chemistry by doing chemistry. No matter what chemistry textbook you use in the classroom, Flinn ChemTopic® Labs will help you give your students the necessary knowledge, skills, attitudes, and values to be successful in chemistry.

About the Curriculum Advisory Board

Flinn Scientific is honored to work with an outstanding group of dedicated chemistry teachers. The members of the Flinn ChemTopic Labs Advisory Board have generously contributed their proven experiments, demonstrations, and teaching tips to create these topic lab manuals. The wisdom, experience, creativity, and insight reflected in their lab activities guarantee that students who perform them will be more successful in learning chemistry. On behalf of all chemistry teachers, we thank the Advisory Board members for their service to the teaching profession and their dedication to the field of chemistry education.

Bob Becker teaches chemistry and AP chemistry at Kirkwood High School in Kirkwood, MO. Bob received his B.A. from Yale University and M.Ed. from Washington University and has 15 years of teaching experience. A well-known demonstrator, Bob has conducted more than 100 demonstration workshops across the U.S. and Canada and is currently a Team Leader for the Flinn Foundation Summer Workshop Program. His creative and unusual demonstrations have been published in the *Journal of Chemical Education,* the *Science Teacher,* and *Chem13 News*. Bob is the author of two books of chemical demonstrations, *Twenty Demonstrations Guaranteed to Knock Your Socks Off, Volumes I and II,* published by Flinn Scientific. Bob has been awarded the James Bryant Conant Award in High School Teaching from the American Chemical Society, the Regional Catalyst Award from the Chemical Manufacturers Association, and the Tandy Technology Scholar Award.

Kathleen J. Dombrink teaches chemistry and advanced-credit college chemistry at McCluer North High School in Florissant, MO. Kathleen received her B.A. in Chemistry from Holy Names College and M.S. in Chemistry from St. Louis University and has more than 30 years of teaching experience. Recognized for her strong support of professional development, Kathleen has been selected to participate in the Fulbright Memorial Fund Teacher Program in Japan and NEWMAST and Dow/NSTA Workshops. She served as co-editor of the inaugural issues of *Chem Matters* and was a Woodrow Wilson National Fellowship Foundation Chemistry Team Member for more than 10 years. Kathleen is currently a Team Leader for the Flinn Foundation Summer Workshop Program. Kathleen has been awarded the Midwest Regional Teaching Award from the American Chemical Society, the Tandy Technology Scholar Award, and a Regional Catalyst Award from the Chemical Manufacturers Association.

Robert Lewis teaches chemistry and AP chemistry at Downers Grove North High School in Downers Grove, IL. Robert received his B.A. from North Central College and M.A. from University of the South and has more than 25 years of teaching experience. He was a founding member of Weird Science, a group of chemistry teachers that has traveled throughout the country to stimulate teacher interest and enthusiasm for using demonstrations to teach science. Robert was a Chemistry Team Leader for the Woodrow Wilson National Fellowship Foundation and is currently a Team Leader for the Flinn Foundation Summer Workshop Program. Robert has received the Presidential Award, the James Bryant Conant Award in High School Teaching from the American Chemical Society, the Tandy Technology Scholar Award, a Regional Catalyst Award from the Chemical Manufacturers Association, and a Golden Apple Award from the State of Illinois.

John G. Little teaches chemistry and AP chemistry at St. Mary's High School in Stockton, CA. John received his B.S. and M.S. in Chemistry from University of the Pacific and has more than 35 years of teaching experience. Highly respected for his well-designed labs, John is the author of two lab manuals, *Chemistry Microscale Laboratory Manual* (DC Heath), and *Microscale Experiments for General Chemistry* (with Kenneth Williamson, Houghton Mifflin). He is also a contributing author to *Science Explorer* (Prentice Hall) and *World of Chemistry* (McDougal Littell). John served as a Chemistry Team Leader for the Woodrow Wilson National Fellowship Foundation from 1988 to 1997 and is currently a Team Leader for the Flinn Foundation Summer Workshop Program. He has been recognized for his dedicated teaching with the Tandy Technology Scholar Award and the Regional Catalyst Award from the Chemical Manufacturers Association.

Lee Marek teaches chemistry and AP chemistry at Naperville North High School in Naperville, IL. Lee received his B.S. in Chemical Engineering from the University of Illinois and M.S. degrees in both Physics and Chemistry from Roosevelt University. He has more than 30 years of teaching experience and is currently a Team Leader for the Flinn Foundation Summer Workshop Program. His students have won national recognition in the International Chemistry Olympiad, the Westinghouse Science Talent Search, and the Internet Science and Technology Fair. Lee was a founding member of ChemWest, a regional chemistry teachers alliance, and led this group for 14 years. Together with two other ChemWest members, Lee also founded Weird Science and has presented 500 demonstration and teaching workshops for more than 300,000 students and teachers across the country. Lee has performed science demonstrations on the *David Letterman Show* 20 times. Lee has received the Presidential Award, the James Bryant Conant Award in High School Teaching from the American Chemical Society, the National Catalyst Award from the Chemical Manufacturers Association, and the Tandy Technology Scholar Award.

John Mauch teaches chemistry and AP chemistry at Braintree High School in Braintree, MA. John received his B.A. in Chemistry from Whitworth College and M.A. in Curriculum and Education from Washington State University and has 25 years of teaching experience. John is an expert in "writing to learn" in the chemistry curriculum and in microscale chemistry. He is the author of two lab manuals, *Chemistry in Microscale, Volumes I and II* (Kendall/Hunt). He is also a dynamic and prolific demonstrator and workshop leader. John has presented the Flinn Scientific Chem Demo Extravaganza show at NSTA conventions for seven years and has conducted more than 100 workshops across the country. John was a Chemistry Team Member for the Woodrow Wilson National Fellowship Foundation program for four years and is currently a Team Leader for the Flinn Foundation Summer Workshop Program.

Dave Tanis is Associate Professor of Chemistry at Grand Valley State University in Allendale, MI. Dave received his B.S. in Physics and Mathematics from Calvin College and M.S. in Chemistry from Case Western Reserve University. He taught high school chemistry for 25 years before joining the staff at Grand Valley State University to direct a coalition for improving precollege math and science education. Dave later joined the faculty at Grand Valley State University and currently teaches courses for pre-service teachers. The author of two laboratory manuals, Dave acknowledges the influence of early encounters with Hubert Alyea, Marge Gardner, Henry Heikkinen, and Bassam Shakhashiri in stimulating his long-standing interest in chemical demonstrations and experiments. Continuing this tradition of mentorship, Dave has led more than 40 one-week institutes for chemistry teachers and served as a Team Member for the Woodrow Wilson National Fellowship Foundation for 13 years. He is currently a Board Member for the Flinn Foundation Summer Workshop Program. Dave received the College Science Teacher of the Year Award from the Michigan Science Teachers Association.

Preface

The Periodic Table

Everything you ever wanted to know about chemistry is in the periodic table! From its discovery more than 125 years ago, the periodic table has served as both the most recognized symbol of chemistry across the world and the most important tool for understanding the chemistry of the elements. The purpose of *The Periodic Table,* Volume 4 in the Flinn ChemTopic® Labs series, is to provide high school teachers with laboratory activities that will allow students to discover for themselves key principles of the periodic law for the classification of elements. *The Periodic Table* is a collection of four experiments and five demonstrations/activities that take students on a guided tour of the periodic table—its discovery, predictions, and applications. Students learn about the discovery of the periodic table, the existence of trends or relationships among the elements that can be used to predict the properties of related or unknown elements, and the organization of elements into groups and periods based on similarities, differences, and trends in their properties.

Analyze and discover

Dmitri Mendeleev's discovery of the periodic table provides an interesting parallel to the struggles students experience as they confront the rapidly expanding body of knowledge in modern chemistry. More than 60 elements were known to Mendeleev, along with a great many facts about their properties. What was missing was a way to organize these facts, a way to understand how individual facts related to each other. "It's in the Cards" is an inquiry-based lab activity that allows students to reenact Mendeleev's discovery of the periodic law. Students analyze known facts about the elements using a special set of element cards that summarize the properties of elements, but not their names or identities. As students arrange and rearrange the element cards based on logical and repeating trends in some of the properties, they discover the nature and importance of the periodic law. "Plotting Trends" is a cooperative activity that allows students to analyze quantities that are too small to be seen or measured directly. Students use miscroscale reaction plates and straws to build three-dimensional models and discover periodic trends.

Classify and predict

Dmitri Mendeleev was not the first scientist to recognize relationships among the elements and classify them into groups. He was, however, the first to predict the existence and properties of missing elements when the elements were arranged in order of increasing atomic mass. In "Density Is a Periodic Property," students re-create the excitement that followed the prediction and discovery of Mendeleev's missing elements. Students measure mass and volume data for three elements, calculate their densities, then graph and analyze the results to predict the density of germanium, Mendeleev's missing element in the Group IV family of elements. "The Ultimate Element Crossword" is a paper activity that tests student knowledge of the elements, their classification, and their uses in modern society.

Recognize and explain

Similarities and differences among the elements give rise to so-called periodic trends, both across rows and within columns of the periodic table. Recognizing periodic trends in the physical and chemical properties of the elements is key to understanding the full value of the periodic table. "Periodic Trends and the Properties of Elements" is a microscale experiment in which students compare the reactions of the alkaline earth metals and identify the trend in metal activity both within a group and across a series. Two complementary demonstrations—"Periodic Activity of Metals" and "Safe Swimming with Sodium"—illustrate the spectacular reactions of the more active alkali metals. In "All in the Family," students investigate the periodic trend in the activity of nonmetals within the halogen group. The "Solubility Patterns" chemical demonstration explores the solubility pattern of alkaline earth metal compounds.

Safety, flexibility and choice

Depend on Flinn Scientific to give you the information and confidence you need to work safely with all chemicals, including reactive metals and halogens. *The Periodic Table* offers true flexibility and choice to meet the needs of a diverse student body in diverse settings. Do you have a group of students who think chemistry means memorizing lots of isolated facts? "It's in the Cards" will challenge these students—there are no right or wrong answers! Two experiments explore similarities and differences within families of elements. "Periodic Trends and the Properties of Elements" is more straightforward; the reactions are easy to see and understand. "All in the Family" is more challenging, because students have to think about what they are seeing and what the observations mean. Students who are visual learners will benefit from the opportunity to build three-dimensional models of periodic trends in "Plotting Trends." Using "The Ultimate Element Crossword Puzzle" means you don't have to leave behind the students who love the language arts. Use the experiment summaries and concepts to locate the concepts you want to teach and to choose experiments and demonstrations that will help you meet your goals.

Format and Features

Flinn ChemTopic® Labs

All experiments and demonstrations in Flinn ChemTopic® Labs are printed in a 10⅞″ × 11″ format with a wide 2″ margin on the inside of each page. This reduces the printed area of each page to a standard 8½″ × 11″ format suitable for copying. The wide margin assures you the entire printed area can be easily reproduced without hurting the binding. The margin also provides a convenient place for teachers to add their own notes.

Concepts

Use these bulleted lists along with state and local standards, lesson plans, and your textbook to identify activities that will allow you to accomplish specific learning goals and objectives.

Background

A balanced source of information for students to understand why they are doing an experiment, what they are doing, and the types of questions the activity is designed to answer. This section is not meant to be exhaustive or to replace the students' textbook, but rather to identify the core concepts that should be covered before starting the lab.

Experiment Overview

Clearly defines the purpose of each experiment and how students will achieve this goal. Performing an experiment without a purpose is like getting travel directions without knowing your destination. It doesn't work, especially if you run into a roadblock and need to take a detour!

Pre-Lab Questions

Making sure that students are prepared for lab is the single most important element of lab safety. Pre-lab questions introduce new ideas or concepts, review key calculations, and reinforce safety recommendations. The pre-lab questions may be assigned as homework in preparation for lab or they may be used as the basis of a cooperative class activity before lab.

Materials

Lists chemical names, formulas, and amounts for all reagents—along with specific glassware and equipment—needed to perform the experiment as written. The material dispensing area is a main source of student delay, congestion, and accidents. Three dispensing stations per room are optimum for a class of 24 students working in pairs. To safely substitute different items for any of the recommended materials, refer to the *Lab Hints* section in each experiment or demonstration.

Safety Precautions

Instruct and warn students of the hazards associated with the materials or procedure and give specific recommendations and precautions to protect students from these hazards. Please review this section with students before beginning each experiment.

Procedure

This section contains a stepwise, easy-to-follow procedure, where each step generally refers to one action item. Contains reminders about safety and recording data where appropriate. For inquiry-based experiments the procedure may restate the experiment objective and give general guidelines for accomplishing this goal.

Data Tables

Data tables are included for each experiment and are referred to in the procedure. These are provided for convenience and to teach students the importance of keeping their data organized in order to analyze it. To encourage more student involvement, many teachers prefer to have students prepare their own data tables. This is an excellent pre-lab preparation activity—it ensures that students have read the procedure and are prepared for lab.

Post-Lab Questions or Data Analysis

This section takes students step-by-step through what they did, what they observed, and what it means. Meaningful questions encourage analysis and promote critical thinking skills. Where students need to perform calculations or graph data to analyze the results, these steps are also laid out sequentially and in order.

Format and Features

Teacher's Notes

Master Materials List

Lists the chemicals, glassware, and equipment needed to perform the experiment. All amounts have been calculated for a class of 30 students working in pairs. For smaller or larger class sizes or different working group sizes, please adjust the amounts proportionately.

Preparation of Solutions

Calculations and procedures are given for preparing all solutions, based on a class size of 30 students working in pairs. With the exception of particularly hazardous materials, the solution amounts generally include 10% extra to account for spillage and waste. Solution volumes may be rounded to convenient glassware sizes (100 mL, 250 mL, 500 mL, etc.)

Safety Precautions

Repeats the safety precautions given to the students and includes more detailed information relating to safety and handling of chemicals and glassware. Refers to Material Safety Data Sheets that should be available for all chemicals used in the laboratory.

Disposal

Refers to the current *Flinn Scientific Catalog/Reference Manual* for general guidelines and specific procedures governing the disposal of laboratory waste. Because we recommend that teachers review local regulations before beginning any disposal procedure, the information given in this section is for general reference purposes only. However, if a disposal step is included as part of the experimental procedure itself, then the specific solutions needed for disposal are described in this section.

Lab Hints

This section reveals common sources of student errors and misconceptions and where students are likely to need help. Identifies the recommended length of time needed to perform each experiment, suggests alternative chemicals and equipment that may be used, and reminds teachers about new techniques (filtration, pipeting, etc.) that should be reviewed prior to lab.

Teaching Tips

This section puts the experiment in perspective so that teachers can judge in more detail how and where a particular experiment will fit into their curriculum. Identifies the working assumptions about what students need to know in order to perform the experiment and answer the questions. Highlights historical background and applications-oriented information that may be of interest to students.

Sample Data

Complete, actual sample data obtained by performing the experiment exactly as written is included for each experiment. Student data will vary.

Answers to All Questions

Representative or typical answers to all questions. Includes sample calculations and graphs for all data analysis questions. Information of special interest to teachers only in this section is identified by the heading *"Note to the teacher."* Student answers will vary.

Look for these icons in the *Experiment Summaries and Concepts* section and in the *Teacher's Notes* of individual experiments to identify inquiry-, microscale-, and technology-based experiments, respectively.

Experiment Summaries and Concepts

Experiment

It's in the Cards—Discovering the Periodic Law

Mendeleev's discovery of the periodic table has survived the test of time and stands to this day as the single most important tool to understand the chemistry of the elements. In this inquiry-based activity, students play chemical "solitaire" using special element card decks to reenact Mendeleev's original discovery of the periodic law. As the cards are arranged and rearranged based on logical trends in the properties of the elements, the nature of the periodic law is revealed.

Inquiry

Concepts

- Periodic law
- Periodic table
- Atomic mass
- Ionization energy

Density Is a Periodic Property—Discovering an Element

After observing trends in the properties of elements when they were arranged in order of increasing atomic mass, Mendeleev made a startling prediction. He predicted the existence and properties of at least three undiscovered elements. In this experiment students measure mass and volume data for silicon, tin, and lead, calculate their densities, and use these results to predict the density of germanium, Mendeleev's "undiscovered" element in the Group IV family of elements.

Concepts

- Periodic law
- Density
- Group IVA elements
- Period number

Periodic Trends and the Properties of Elements—Alkaline Earth Metals

Recognizing periodic trends in the properties of elements is key to understanding the full value of the periodic table. The purpose of this microscale lab activity is to identify periodic trends in the activity and solubility of the alkaline earth metals. Students first compare the reactions of metals to discover the trend in metal activity both within a group and across a period in the periodic table. Students then explore the solubility of a variety of alkaline earth metal compounds and use the resulting trend to identify an unknown.

Microscale

Concepts

- Periodic table
- Periodic trends
- Group vs. period
- Alkaline earth metals
- Activity series

All in the Family—The Halogens and Their Compounds

In every family there are both similarities and differences. The same is true in chemical families as well. In this experiment students explore the similarities and differences in the chemical properties of the halogens and their compounds. Students determine which halogens are most and least active by investigating the pair-wise reactions of chlorine, bromine, and iodine with sodium chloride, sodium bromide, and sodium iodide. Students also examine the solubility of metal halides to see how they are the same and different. The experiment is carried out on the microscale level using small test tubes.

Microscale

Concepts

- Periodic table
- Periodic trends
- Groups or families
- Halogens
- Activity series

Experiment Summaries and Concepts

Demonstration

Concepts

Periodic Activity of Metals Chemical Demonstration

The elements are classified based on similarities, differences, and trends in their properties. The reactions of alkali and alkaline earth metals with water are some of the most dramatic of all chemical reactions. Mixtures bubble and boil, hiss and fizz, and may even smoke and burn. Use this exciting demonstration of the activity of metals to introduce the study of the periodic table and periodic trends.

- Periodic trends
- Alkali and alkaline earth metals
- Metal activity
- Physical and chemical properties

Safe Swimming with Sodium Chemical Demonstration

No chemistry class is complete without the spectacular demonstration of the reaction of sodium with water. This novel variation is much safer to perform than the traditional reaction and demonstrates the reaction of sodium in slow motion. In a two-phase system composed of mineral oil and water, sodium sits at the interface between the two layers. Sodium reacts with water, picks up hydrogen gas bubbles, and floats to the oil surface. There it gradually loses the gas bubbles, then dives back down to the water layer and reacts again. Students are fascinated as the process repeats itself over and over again.

- Alkali metals
- Metal hydroxides
- Acid–base indicator
- Density

Plotting Trends—A Periodic Table Activity

Ionization energies, atomic radii, electronegativity values—what do all these numbers mean? Students cannot measure them, cannot see their relative size. In this cooperative lab activity, students use microscale reaction plates and straws of different lengths to construct three-dimensional bar charts of the physical properties of the elements. The resulting charts are visually impressive and clearly illustrate the existence, direction, and meaning of periodic trends.

- Periodic table
- Periodic trends
- Ionization energy
- Atomic radius
- Electronegativity

The Ultimate Element Crossword—A Periodic Table Activity

What alkaline earth metal is very toxic, yet given to patients as a nice thick sulfate shake so that doctors won't have to *bury them?* Barium! Increase student knowledge of the elements, their uses, and their position in the periodic table with this interesting and entertaining crossword puzzle. Many of the puzzle clues contain fun "plays on words" that, while they may sound *so dumb,* will make students want to find them. Sodium!

- Elements
- Descriptive chemistry
- Groups or families
- Periods or series

Solubility Patterns Chemical Demonstration

The solubility of alkaline earth metal compounds varies widely, depending on the metal cation and the nature of the anion. Are there any patterns or trends in the solubility behavior of alkaline earth metal compounds? If you give your students the answer to this question, they may remember it only long enough to repeat it on an exam. Let students discover the pattern for themselves, however, and they are likely not only to remember it long after, but to explain it as well.

Microscale

- Alkaline earth metals
- Periodic trends
- Double replacement reactions
- Ionic compounds

Teacher Notes

It's in the Cards
Discovering the Periodic Law

Introduction

Dmitri Mendeleev's discovery of the Periodic Law ranks as one of the greatest achievements in the history of science. It has survived the test of time and stands to this day as the single most important tool to understand the chemistry of the elements. As we try to understand the essence of this great discovery, it is worthwhile to go back in time and look at how it was achieved. What did Mendeleev know and when did he know it?

Concepts

- Periodic law
- Periodic table
- Atomic mass
- Ionization energy

Background

In the years 1868–1870, Dmitri Mendeleev, a professor of chemistry at the University of St. Petersburg in Russia, was intent on writing a new textbook of chemistry, called *Principles of Chemistry*. More than 60 individual elements were known, and a great many facts had been discovered about their properties and compounds. Mendeleev knew the atomic masses of the elements, their densities, boiling points, melting points, the formulas of their compounds with hydrogen, oxygen, and chlorine, and many other isolated facts. What was missing was a way to organize these facts, a way to understand how individual facts related to each other—in short, a way to classify the elements. The following quote is translated from Mendeleev's first published report of the periodic law:

> *"I wished to establish some sort of system of elements in which their distribution is not guided by chance . . . but by some sort of definite and exact principle."*

Mendeleev decided to arrange the elements systematically according to their atomic mass. He wrote out the exact atomic masses (as they were known at the time) in the margin of a list of the elements, then wrote out separate cards for each of the elements, with their atomic mass and important chemical and physical properties. Using these cards, Mendeleev played "chemical solitaire" for several hours, finally copying to a sheet of paper the arrangement he had worked out with the cards. With slight modification, this became Mendeleev's first *Periodic Table of the Elements*.

The meaning of the term "periodic" is not obvious to most students. Use some real-life examples to explain its meaning. Magazines are called periodicals because they are published periodically—weekly, monthly, semiannually, etc. Some other periodic functions in our daily lives are the days of the week, the phases of the moon, and the changing of the seasons.

Experiment Overview

The purpose of this activity is to re-create Mendeleev's discovery of the classification of the elements and the periodic law using a special deck of element cards. The real properties of the elements, but not their names or symbols, are written on these cards. As the cards are arranged and rearranged based on logical trends in some of these properties, the nature of the periodic law should reveal itself.

Teacher Notes

Pre-Lab Questions

The element cards list all of the following properties for each element. Use your textbook to write a clear definition of each property. Give the typical units for each numerical property, if appropriate.

- Ionization energy
- Atomic radius
- Melting point
- Atomic mass
- Density
- Electronegativity

Materials

Special deck of 31 element cards

Procedure

1. Form a research group with two or three other students. Obtain a deck of element cards and spread the cards out on the lab table.
2. Each card lists the properties of a single element (X), as shown below:

Ionization energy	**Atomic radius**
Atomic mass	**Formula of its oxide X_bO_c**
Formula of its chloride XCl_a	**Melting point**
Density*	**Formula of its hydride XH_d**
Electronegativity	

*Density values are in units of g/cm^3 for solids and liquids, g/L for gases.

3. Working in a collaborative manner, discuss the possibilities for arrangement of the element cards with all members of the group, and look for a logical arrangement of the cards. Consider the following questions:
   - What are the similarities and differences among the elements?
   - Are there any numerical or logical trends in their properties?
4. It is NOT within the rules of this game of chemical solitaire to look up information in a textbook or to use a modern periodic table as a guide!

Mendeleev's greatest insight in creating the periodic table was in recognizing there were some gaps when the elements were arranged in logical order. He had the courage not only to leave blanks in his table for the missing elements, but also to predict their properties.

Teacher Notes

5. One of the element cards is also missing in your deck of cards. Decide where the missing element belongs in the arrangement of the elements and rearrange the cards if necessary to accommodate the missing element.

6. On the Data Sheet, fill in the *Table of the Elements* to illustrate a logical arrangement of the element cards. To do this, write down only the *atomic mass* of each element, as shown on its card. Leave a blank space for the missing element. *Note:* The table is 8 × 8 and contains 32 more squares than are needed—plenty of room to arrange the elements many different ways.

7. Predict the properties of the missing element by averaging the properties of its nearest neighbors (in any direction). On the Data Sheet, complete the card for the missing element by entering its predicted properties alongside the name of each property.

Teacher Notes

Name: ____________________

Class/Lab Period: ____________________

It's in the Cards

Data Sheet

Table of the Elements

Properties of the Missing Element

Ionization energy

Atomic mass

Formula of its chloride

Density

Electronegativity

Atomic radius

Formula of its oxide

Melting point

Formula of its hydride

Teacher Notes

Post-Lab Questions *(Use a separate sheet of paper to answer the following questions.)*

1. Mendeleev's Periodic Law can be stated: *"The physical and chemical properties of elements are periodic functions of their atomic masses".* Looking at your arrangement of the element cards, describe in your own words what the term "periodic function" means.
2. Some of the the properties listed on each card are periodic properties, others are not. Name one property that is periodic and one that is not.
3. The elements in the modern periodic table are arranged in order of increasing atomic number (instead of increasing atomic mass). Why didn't Mendeleev use atomic number to arrange the elements?
4. From your instructor, obtain a handout showing one possible arrangement of the element cards. Identify each of the elements on the handout with its *atomic number* and *chemical symbol*.
5. Using the possible arrangement of the element cards obtained from your instructor, pick two of the *numerical properties* of the elements that are periodic and plot their values on the graphs below. Give each graph a descriptive title and label the axes.

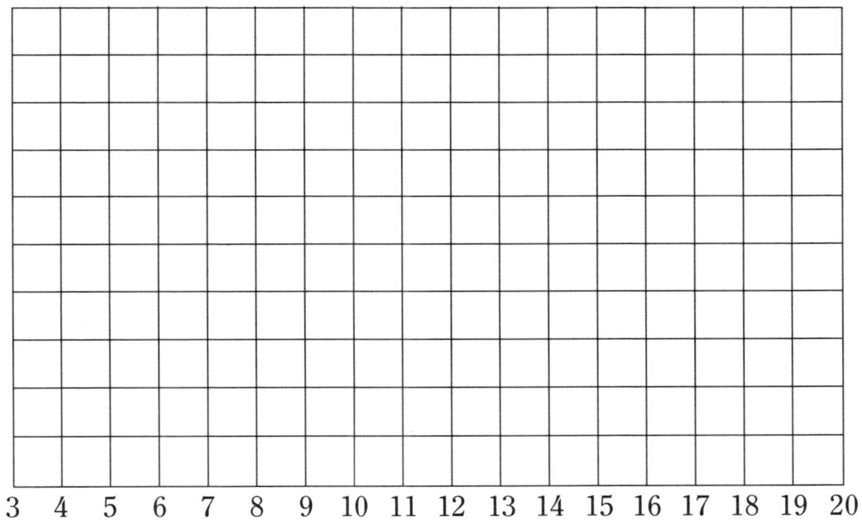

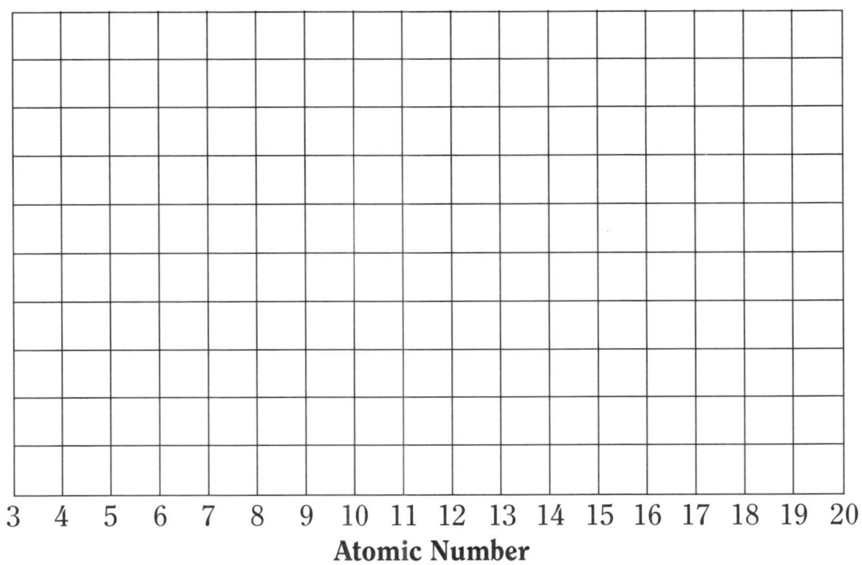

Teacher Notes

Post-Lab Questions (continued)

6. There are certain trends in the properties of the elements, both within a column (from top to bottom) and across a row (from left to right) in the periodic table. On the arrow for each property, write the word *increases* or *decreases* to describe how that property changes.

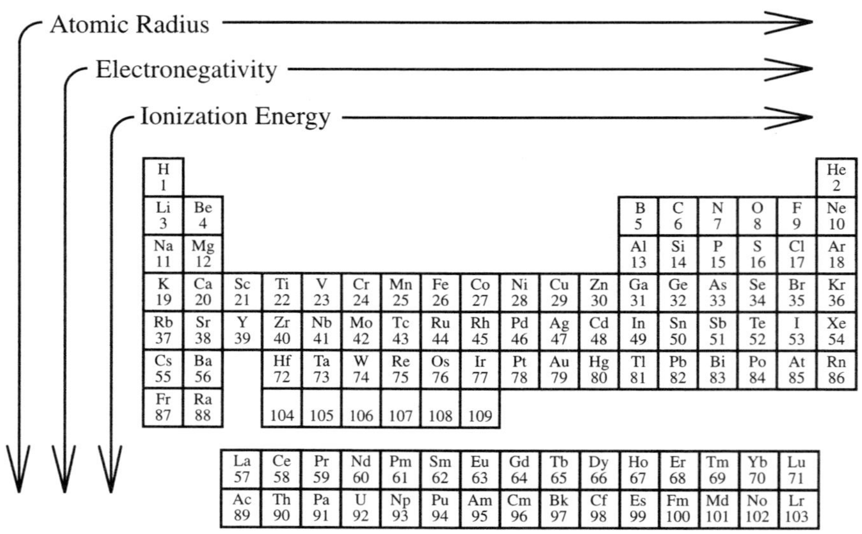

7. On the outline of the periodic table shown below, locate the following: *metals, nonmetals,* and *metalloids* (or *semimetals*). Use your textbook to define these terms, if necessary.

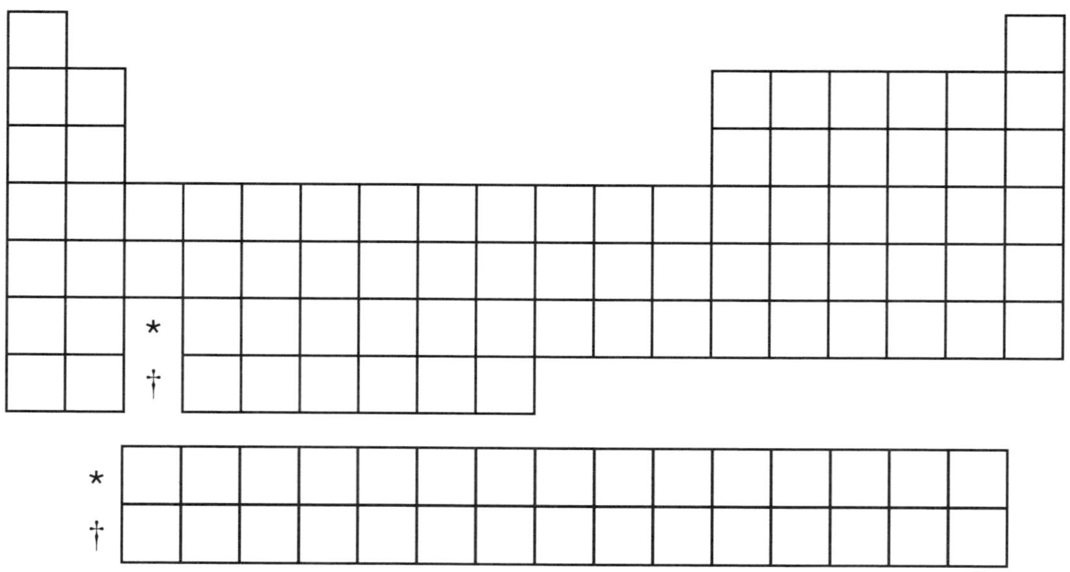

Teacher Notes

Post-Lab Questions (continued)

8. On the outline of the periodic table shown below, locate the following: *groups or families of elements, periods or series of elements, noble gases, alkali metals, alkaline earth metals, and halogens*. Use your textbook to define these terms, if necessary.

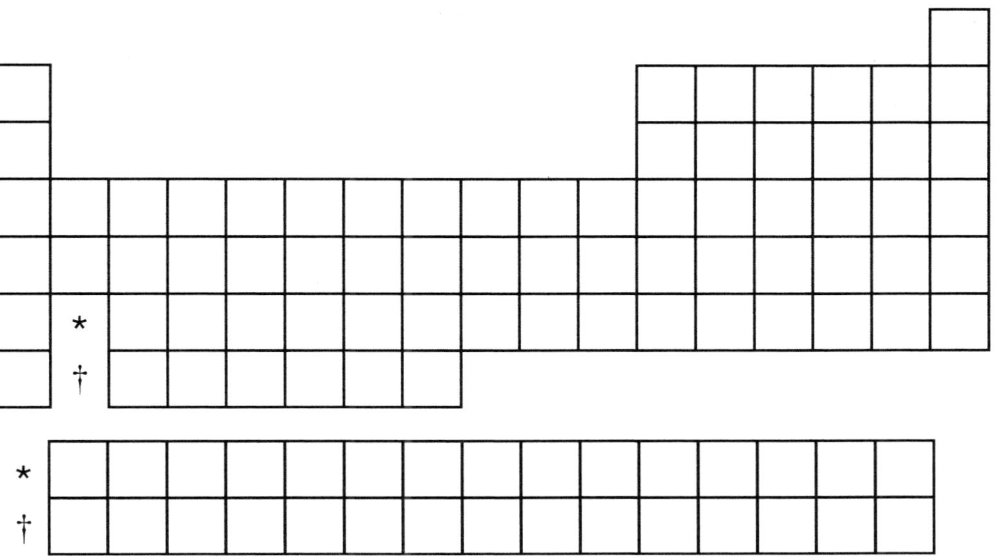

9. On the outline of the Periodic Table shown below, locate the following: *transition elements, inner transition elements, representative elements*. Use your textbook to define these terms, if necessary.

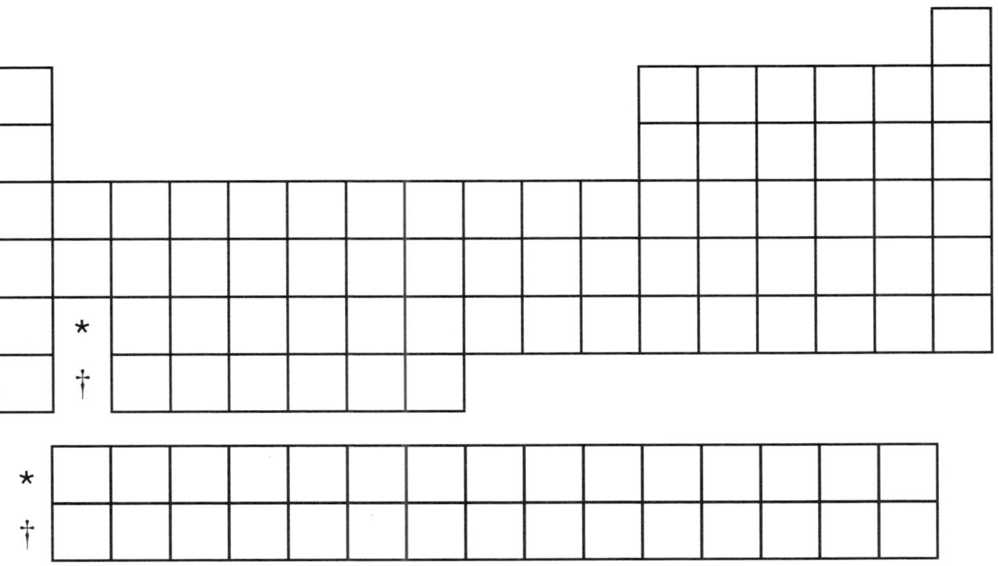

There is still some inconsistency in the naming of the representative and inner transition elements. The representative elements were at one time referred to as the main group elements, and the inner transition elements as the lanthanides and actinides.

It's in the Cards

Inquiry

Master Materials List *(for a class of 30 students working in groups of 3–4)*

Special deck of element cards, 7–10 sets

Handout showing possible arrangement of the element cards, 30

Materials Preparation

Element Card Decks: Make copies of the element cards (see the Supplementary Information section, pages 15–16) and laminate the sheets. Following the dotted lines, cut out the individual element boxes to obtain one "deck" of element cards. Shuffle each deck of cards, or have the students shuffle them, prior to beginning the exercise. To improve the longevity of the element card deck, copy the handout onto heavy card stock prior to laminating it.

Element Card Arrangement Handout: Make enough copies of this handout on regular paper to give each student an individual copy. The students will work from these handouts to complete the post-lab questions and graphing exercises.

Lab Hints

- Although this is a "dry-lab" activity, it is best to schedule enough time for it in class so that students can complete Steps 1–6 in the Procedure section. This will ensure that students really do treat the lab as a discovery activity. Do not give students the handout with the suggested element arrangement until they have completed the *Table of the Elements* on the Data Sheet. Some teachers go even further and require their students to "commit" to their arrangement in ink before giving them the handout. The post-lab questions can be assigned as homework or as part of a more formal lab report.

- Keeping in mind the spirit of discovery, remind students that there is no single correct answer for the arrangement of the elements. Students should not be graded, therefore, on the "correctness" of their Table of the Elements, but rather on what they learn in the process. Remember that in Mendeleev's first published table of the elements (1869), elements were listed in order of increasing atomic mass in columns, so that similar elements (e.g., Cl, Br, I) were found in horizontal rows rather than vertical columns. The vertical arrangement of similar elements into element groups appeared two years later, in 1871.

Teaching Tips

- Although Dmitri Mendeleev is often given top billing in textbook discussions of the development of the periodic table, Lothar Meyer in Germany is credited with independent discovery of the periodic law. In an 1870 paper, Meyer printed a graph of atomic volume versus atomic weight that revealed the characteristic repeating pattern in the properties of elements. In the same paper, Meyer also displayed an arrangment of the elements into a table of rows and columns on the basis of increasing atomic weight. Meyer did not, however, predict the existence of missing or undiscovered elements, as Mendeleev did.

Teacher Notes

Will the card decks be used by several classes in a day? You may want to number or label the cards in some way to quickly identify missing cards. Another suggestion: Copy separate card decks on different color paper to make it easy to sort the decks at the end of class. Always have students shuffle the cards before beginning the activity.

Teacher Notes

- Both Mendeleev and Meyer built on the work of earlier scientists, notably that of Döbereiner and Newlands. In 1829, Döbereiner classified similar elements into triads, or groups of three, in which the atomic weight of the middle element was the average of the other two elements. Some of his triads included now-familiar families of elements: lithium, sodium, and potassium; calcium, strontium, and barium; sulfur, selenium, and tellurium; and chlorine, bromine, and iodine. In 1865, Newlands published an eight-column table of the elements, in which 62 elements were arranged in order of both increasing atomic weight and properties that repeated themselves every eight elements.

Answers to Pre-Lab Questions

1. **Ionization energy** — energy needed to remove one of the electrons from an atom (specifically, from an isolated atom in the gas phase). The energy needed to remove successive electrons is called the first ionization energy, second ionization energy, etc. Units are kilojoules per mole (kJ/mole).

2. **Atomic radius** — for a neutral atom, the distance from the center of an atom's nucleus to its outermost electron. Since the electron cloud that surrounds the nucleus does not have an exact boundary, the atomic radius of an atom is not a precise number. The atomic radius is usually estimated as one-half the distance between two atoms in a form of the element—for example, one-half the distance between chlorine atoms in the diatomic chlorine molecule (Cl_2) or between two carbon atoms in diamond, the most dense allotrope of carbon. Units are nanometers ($1 \text{ nm} = 1 \times 10^{-9} \text{ m}$).

3. **Density** — the ratio of mass to unit volume for the standard form of the element. Units are grams per cubic centimeter (g/cm^3) for a solid or liquid and grams per liter (g/L) for a gas. Density depends on temperature.

4. **Electronegativity** — assigned numerical value for an element that reflects the ability of an atom to attract electron density in a covalent bond. Electronegativity is not an amount of energy, nor can it be directly measured. There are no units for electronegativity; it is assigned on a relative scale from approximately 0.5 to 4.0.

5. **Atomic mass** — also called the average atomic mass; the average mass of an element's atoms. Since the mass of an individual atom is too small to be directly measured, atomic mass is defined relative to that of a reference atom, an arbitrary standard. Thus, the mass of the carbon-12 isotope is defined as exactly equal to 12 atomic mass units. The atomic mass of a particular atom is very nearly, but not exactly, equal to its mass number (number of protons and neutrons). The average atomic mass of an element is computed as a weighted average of the masses of the isotopes that make up the element and their fractional (natural) abundance. The units of atomic mass are atomic mass units (amu).

6. **Melting point** — temperature at which the solid form of an element is in equilibrium with the liquid phase at atmospheric pressure. Usually given in units of kelvin (K).

An interesting extension of this activity (extra credit?) would be to have students research the dates of discovery of the elements. They could then construct a "Periodic Table of Discovery." Are there any patterns or trends in the discovery dates?

Teacher Notes

Sample Data

"Vertical" Table of the Elements

6.9	9.0	10.8	12.0	14.0	16.0	19.0	20.2
23.0	24.3	27.0	28.1	31.0	32.1	35.5	39.9
39.1	40.1	69.7	72.6	74.9	Missing Element	79.9	83.8
85.5	87.6	114.8	118.7	121.8	127.6	126.9	131.3

"Horizontal" Table of the Elements

6.9	23.0	39.1	85.5				
9.0	24.3	40.1	87.6				
10.8	27.0	69.7	114.8				
12.0	28.1	72.6	118.7				
14.0	31.0	74.9	121.8				
16.0	32.1	Missing Element	127.6				
19.0	35.5	79.9	126.9				
20.2	39.9	83.8	131.3				

Notice the inversion of atomic masses between Ar (element 18, atomic mass 39.9) and K (element 19, atomic mass 39.1), and also between Te (element 52, atomic mass 127.6) and I (element 53, atomic mass 126.9). Mendeleev noticed the inversion between Te and I and predicted that the atomic masses were in error. (They were not.) The noble gases (including Ar) were not known at the time of Mendeleev.

Teacher Notes

Properties of the Missing Element

Ionization energy 950	**Atomic radius** 0.12
Atomic mass 77.4	**Formula of its oxide** XO_2
Formula of its chloride XCl_2	**Melting point** 600
Density 4.4	**Formula of its hydride** XH_2
Electronegativity 2.6	

Answers to Post-Lab Questions *(Students answers will vary.)*

1. Mendeleev's Periodic Law can be stated: *"The physical and chemical properties of elements are periodic functions of their atomic masses."* Looking at your arrangement of the element cards, describe in your own words what the term "periodic function" means.

   "Periodic function" means that the properties of the elements repeat in a regular manner (are periodic) and that the cycle for the repeating pattern depends on (is a function of) the atomic masses of the elements.

2. Some of the the properties listed on each card are periodic properties, others are not. Name one property that is periodic and one that is not.

   Ionization energy, electronegativity, and atomic radius reveal distinct periodic trends when the elements are arranged in order of increasing atomic mass. Density and melting point show trends, but they are not as obvious. Atomic mass is not a periodic property.

3. The elements in the modern periodic table are arranged in order of increasing atomic number (instead of increasing atomic mass). Why didn't Mendeleev use atomic number to arrange the elements?

   The concept of atomic number did not exist at the time of Mendeleev. In 1913, the English chemist Henry Moseley, working in Ernest Rutherford's laboratory on the properties of the nucleus, hypothesized that every element had a different amount of positive charge, and that the amount of positive charge occurred in whole-number increments. Moseley introduced the concept of atomic number to account for the positive charge of the nucleus.

4. From your instructor, obtain a handout showing one possible arrangement of the element cards. Identify each of the elements on the handout with its *atomic number* and *chemical symbol*.

   The elements are listed in order from left to right on page 17, in rows according to their periodic number. The transition metals are not included. The atomic numbers of the elements listed are 3–20; 31–38; and 49–54.

The properties of the missing element shown above are not the real properties. They are estimated properties obtained by averaging the properties of its nearest neighbors and examining the trend within its column and row.

5. Using the possible arrangement of the element cards obtained from your instructor, pick two of the *numerical properties* of the elements that are periodic and plot their values on the graphs below. Give each graph a descriptive title and label the axes.

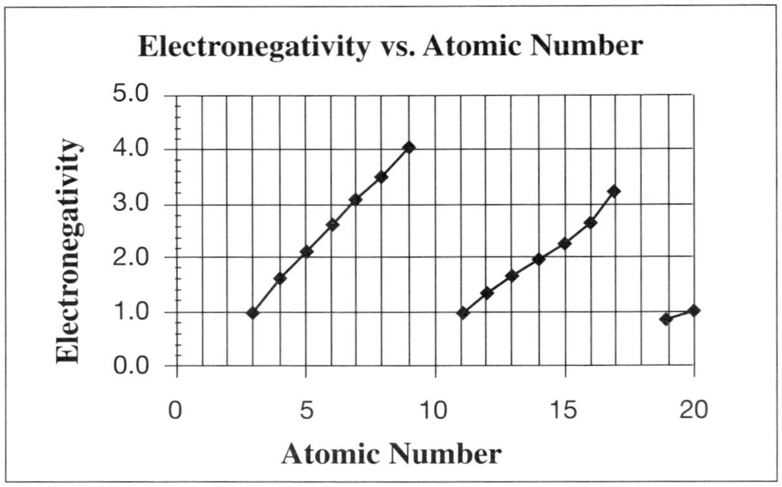

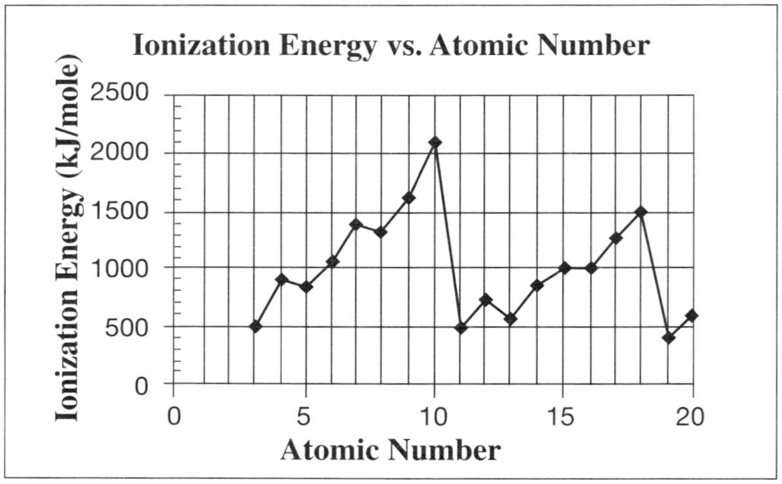

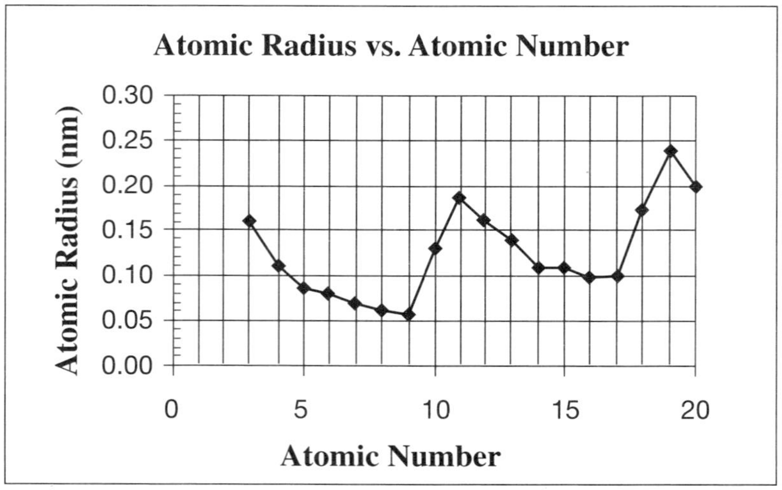

Teacher Notes

Many students will mistakenly draw graphs of atomic mass vs. atomic number to reflect a periodic trend. Atomic mass is not a periodic property. It does not repeat in a regular pattern, but rather steadily increases.

Teacher Notes

6. There are certain trends in the properties of the elements, both within a column (from top to bottom) and across a row (from left to right) in the periodic table. On the arrow for each property, write the word increases or decreases to describe how that property changes.

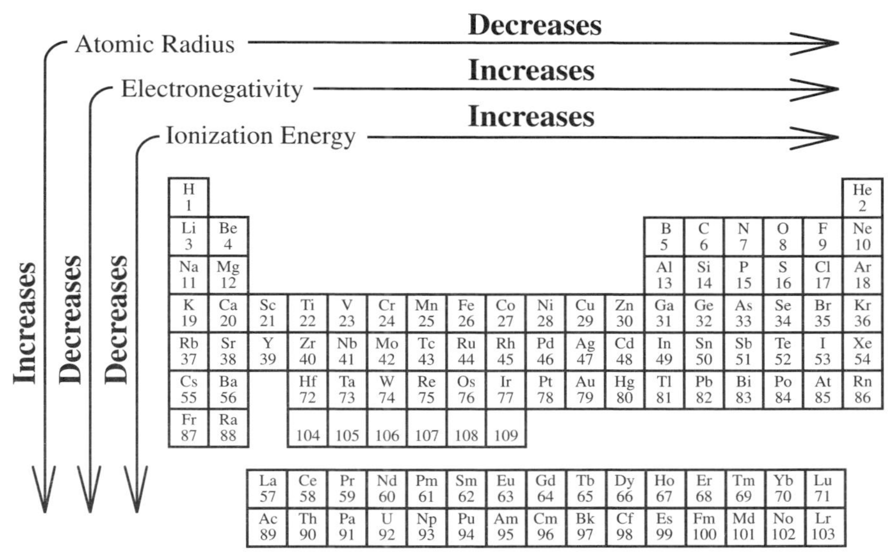

7. On the outline of the periodic table shown below, locate the following: *metals, non-metals,* and *metalloids* (or *semimetals*).

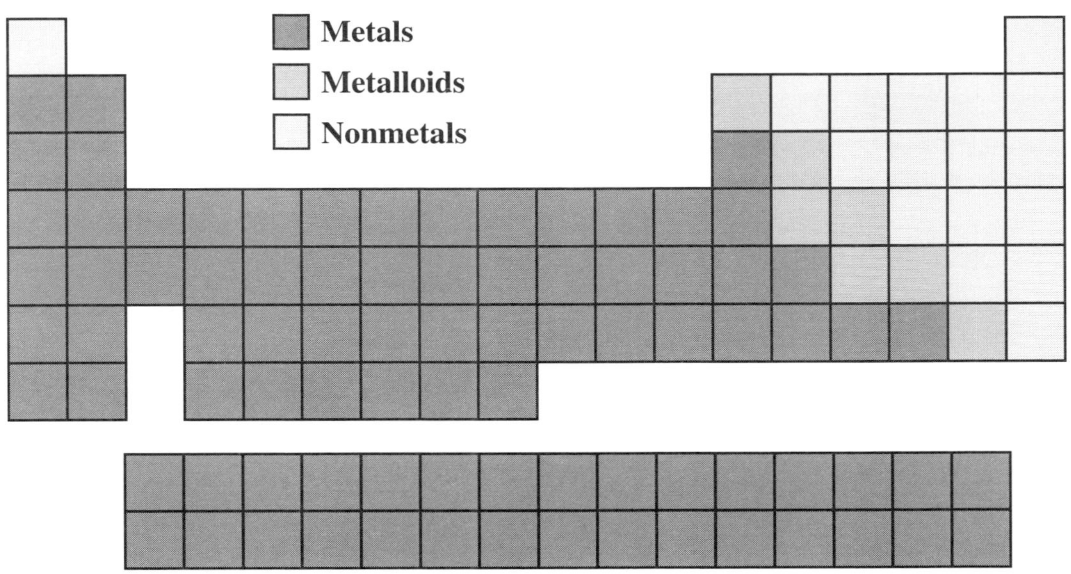

8. On the outline of the periodic table shown below, locate the following: *groups or families of elements, periods or series of elements, noble gases, alkali metals, alkaline earth metals,* and *halogens*.

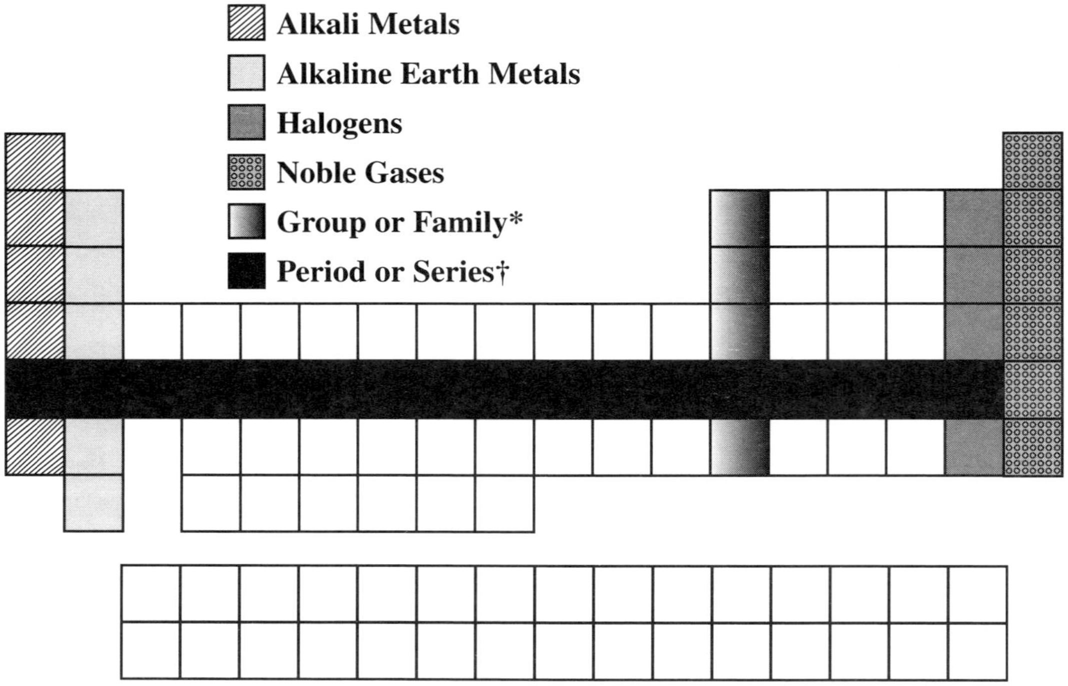

***Every Vertical Column Is a Group**

†Every Horizontal Row Is a Period

9. On the outline of the Periodic Table shown below, locate the following: *transition elements, inner transition elements, representative elements.*

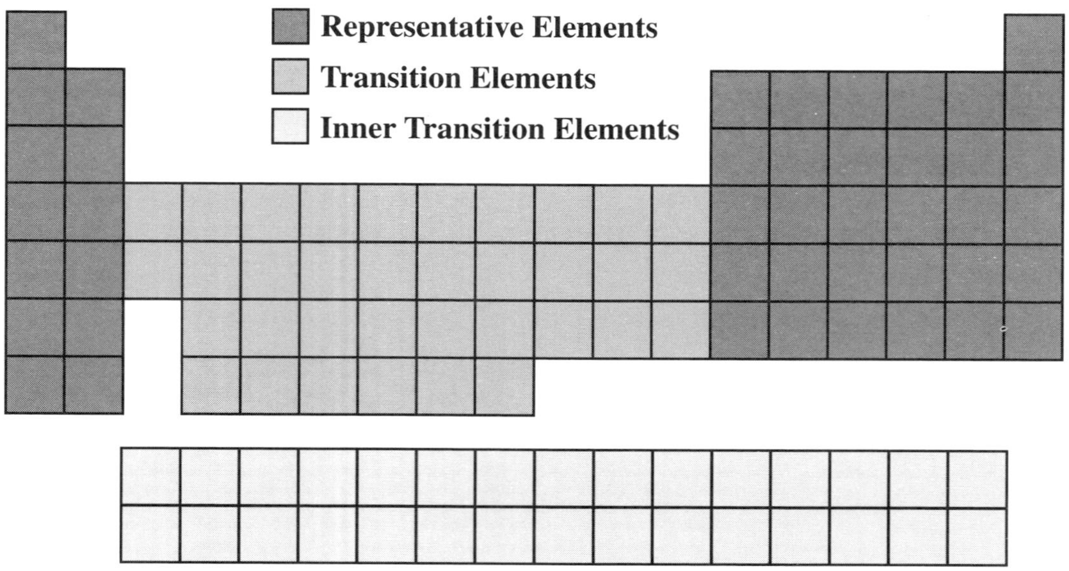

Teacher Notes

Teacher Notes

To create element card decks: Make copies of pages 15 and 16 and laminate the sheets, if desired. Cut out the cards along the dotted lines to obtain one deck of 31 cards (32 if the "missing element" card is included.)

520	0.16
6.9	X_2O
XCl	453.7
0.53	XH
0.98	

899	0.11
9.0	XO
XCl_2	1560
1.85	XH_2
1.57	

801	.083
10.8	X_2O_3
XCl_3	2300
2.34	XH_3
2.04	

1086	.077
12.0	XO_2
XCl_4	4100
2.62	XH_4
2.55	

496	0.19
23.0	X_2O
XCl	371.0
0.97	XH
0.93	

738	0.16
24.3	XO
XCl_2	922
1.74	XH_2
1.31	

578	0.14
27.0	X_2O_3
XCl_3	933.3
2.7	XH_3
1.61	

786	0.11
28.1	XO_2
XCl_4	1685
2.33	XH_4
1.90	

419	0.24
39.1	X_2O
XCl	336.4
0.86	XH
0.82	

590	0.20
40.1	XO
XCl_2	1112
1.55	XH_2
1.00	

579	0.14
69.7	X_2O_3
XCl_3	302.9
5.91	XH_3
1.81	

762	0.12
72.6	XO_2
XCl_4	1210
5.3	XH_4
2.01	

403	0.25
85.5	X_2O
XCl	312.0
1.53	XH
0.80	

549	0.22
87.6	XO
XCl_2	1041
2.6	XH_2
0.95	

558	0.17
114.8	X_2O_3
XCl_3	429.8
7.31	XH_3
1.78	

709	0.16
118.7	XO_2
XCl_4	505.1
7.30	XH_4
1.96	

1402	.070	1314	.066	1681	.062	2081	0.13
14.0	X_2O_5	16.0	XO	19.0	X_2O	20.2	—
XCl_3	63.14	XCl_2	50.35	XCl	53.48	—	24.55
1.25	XH_3	1.4	XH_2	1.7	XH	0.901	—
3.04		3.44		3.98		—	
1012	0.11	1000	0.10	1251	0.10	1521	0.17
31.0	X_2O_5	32.1	XO_2	35.5	X_2O	39.9	—
XCl_3	317.3	XCl_2	388.4	XCl	172.2	—	83.81
1.8	XH_3	2.07	XH_2	3.17	XH	1.78	—
2.19		2.58		3.16		—	
947	0.12			1140	0.12	1351	0.19
74.9	X_2O_5			79.9	X_2O	83.8	—
XCl_3	1081	"Missing Element"		XCl	265.9	—	115.8
5.72	XH_3			3.1	XH	3.74	—
2.18				2.96		—	
834	0.14	869	0.14	1008	0.14	1170	0.21
121.8	X_2O_5	127.6	XO_2	126.9	X_2O	131.3	XO_3
XCl_3	904	XCl_2	722.7	XCl	386.7	—	161.4
6.68	XH_3	6.24	XH_2	4.92	XH	5.89	—
2.05		2.10		2.66		—	

Teacher Notes

Atomic radii are the least well-defined of these characteristic physical properties of the elements. At least three definitions of atomic radius are in common use: covalent radius, Van der Waals radius, and metallic radius. The atomic radius values given here are the metallic radius for all metals and the covalent radius for all nonmetals. All values except those of the Noble Gases are taken from a database of atomic radii maintained with information from N. W. Alcock, Structure and Bonding *(Ellis Horwood, 1990).*

Element Card Arrangement Handout*

<table>
<tr>
<td>520 0.16
6.9 X_2O
XCl 453.7
0.53 XH
0.98</td>
<td>899 0.11
9.0 XO
XCl_2 1560
1.85 XH_2
1.57</td>
<td>801 .083
10.8 X_2O_3
XCl_3 2300
2.34 XH_3
2.04</td>
<td>1086 .077
12.0 XO_2
XCl_4 4100
2.62 XH_4
2.55</td>
<td>1402 .070
14.0 X_2O_5
XCl_3 63.14
1.25 XH_3
3.04</td>
<td>1314 .066
16.0 XO
XCl_2 50.35
1.4 XH_2
3.44</td>
<td>1681 .062
19.0 X_2O
XCl 53.48
1.7 XH
3.98</td>
<td>2081 0.13
20.2 —
— 24.55
0.901 —
—</td>
</tr>
<tr>
<td>496 0.19
23.0 X_2O
XCl 371.0
0.97 XH
0.93</td>
<td>738 0.16
24.3 XO
XCl_2 922
1.74 XH_2
1.31</td>
<td>578 0.14
27.0 X_2O_3
XCl_3 933.3
2.7 XH_3
1.61</td>
<td>786 0.11
28.1 XO_2
XCl_4 1685
2.33 XH_4
1.90</td>
<td>1012 0.11
31.0 X_2O_5
XCl_3 317.3
1.8 XH_3
2.19</td>
<td>1000 0.10
32.1 XO_2
XCl_2 388.4
2.07 XH_2
2.58</td>
<td>1251 0.10
35.5 X_2O
XCl 172.2
3.17 XH
3.16</td>
<td>1521 0.17
39.9 —
— 83.81
1.78 —
—</td>
</tr>
<tr>
<td>419 0.24
39.1 X_2O
XCl 336.4
0.86 XH
0.82</td>
<td>590 0.20
40.1 XO
XCl_2 1112
1.55 XH_2
1.00</td>
<td>579 0.14
69.7 X_2O_3
XCl_3 302.9
5.91 XH_3
1.81</td>
<td>762 0.12
72.6 XO_2
XCl_4 1210
5.3 XH_4
2.01</td>
<td>947 0.12
74.9 X_2O_5
XCl_3 1081
5.72 XH_3
2.18</td>
<td>"Missing Element"</td>
<td>1140 0.12
79.9 X_2O
XCl 265.9
3.1 XH
2.96</td>
<td>1351 0.19
83.8 —
— 115.8
3.74 —
—</td>
</tr>
<tr>
<td>403 0.25
85.5 X_2O
XCl 312.0
1.53 XH
0.80</td>
<td>549 0.22
87.6 XO
XCl_2 1041
2.6 XH_2
0.95</td>
<td>558 0.17
114.8 X_2O_3
XCl_3 429.8
7.31 XH_3
1.78</td>
<td>709 0.16
118.7 XO_2
XCl_4 505.1
7.30 XH_4
1.96</td>
<td>834 0.14
121.8 X_2O_5
XCl_3 904
6.68 XH_3
2.05</td>
<td>869 0.14
127.6 XO_2
XCl_2 722.7
6.24 XH_2
2.10</td>
<td>1008 0.14
126.9 X_2O
XCl 386.7
4.92 XH
2.66</td>
<td>1170 0.21
131.3 XO_3
— 161.4
5.89 —
—</td>
</tr>
</table>

*Represents one possible arrangement.

Key

<table>
<tr><td>Ionization Energy</td><td>Atomic Radius</td></tr>
<tr><td>Atomic Mass</td><td>X_bO_c</td></tr>
<tr><td>XCl_a</td><td>Melting Point</td></tr>
<tr><td>Density†</td><td>XH_d</td></tr>
<tr><td>Electro-negativity</td><td></td></tr>
</table>

†*Density values are in units of g/cm^3 for solids and liquids and g/L for gases.*

Teacher Notes

Teacher Notes

Density Is a Periodic Property
Discovering an Element

Introduction

"With the periodic and atomic relations now shown to exist between all atoms and the properties of their elements, we see the possibility not only of noting the absence of some of them but even of determining . . . properties of these as yet unknown elements."

Dmitri Mendeleev proposed the periodic law for the classification of elements in 1869–1871. After observing trends in the properties of elements when they were arranged in order of increasing atomic mass, Mendeleev made a startling prediction. He predicted the existence and properties of at least three undiscovered elements. Mendeleev saw what other scientists before him had missed—he saw what wasn't there!

Concepts

- Periodic law
- Group IV elements
- Density
- Period number

Background

At the time Mendeleev proposed the periodic law, the foundation of the modern periodic table for the classification of elements, 63 elements were known. Their physical and chemical properties had been studied and their atomic masses measured. Mendeleev arranged the known elements in a calendar-like table of rows and columns in order of increasing atomic mass and repeating chemical properties. It is at this point, however, that Mendeleev made a giant leap of discovery—he suggested that there were some gaps or missing elements in the list of known elements.

Among the Group IV elements in Mendeleev's classification scheme, carbon appeared in the second row, followed by silicon in the third row. Both tin and lead shared similar chemical properties with carbon and silicon and were also known at this time. Because of their high atomic masses, however, these metals were placed in later rows of Mendeleev's Group IV column of elements. In 1871, Mendeleev proposed that there existed an as-yet-unknown element beneath silicon in the Group IV elements. He named the missing element *eka*-silicon and predicted its physical properties (atomic mass, melting point, density, and specific heat). In 1886 the element germanium was discovered by the German chemist Clemens Winkler. In his report of the discovery, Winkler stated: "*. . . There can be no longer any doubt that the new element is no other than the **eka**-silicon prognosticated fifteen years ago by Mendeleev."*

Within 15 years of Mendeleev's prediction of the existence of missing elements, three of the elements had been discovered, their properties in excellent agreement with those predicted by Mendeleev. Is it possible to recreate some of the excitement that followed the prediction and discovery of Mendeleev's missing elements?

Teacher Notes

Experiment Overview

The purpose of this experiment is to measure mass and volume data for silicon, tin, and lead, calculate their densities, and use these results to predict the density of germanium, Mendeleev's "undiscovered" element in the Group IV family of elements. The volume of the elements will be measured by water displacement.

Pre-Lab Questions

1. One of the elements Mendeleev predicted was *eka*-aluminum, corresponding to a gap in the fourth row or period of the Group IIIA elements, between aluminum and indium. The density of aluminum (period 3) is 2.70, that of indium (period 5) 7.31, and that of thallium (period 6) 11.85 g/cm^3. Make a graph of period number on the x-axis versus density on the y-axis for each of these elements.

2. Use your graph to predict the density of *eka*-aluminum. What known element in the modern Periodic Table corresponds to *eka*-aluminum? Look up the density of the modern element in a reference source and record its actual and predicted density values.

3. How do the actual and predicted density values compare? Use the following equation to calculate the percent error between the predicted and actual values for the density of *eka*-aluminum.

$$\text{percent error} = \frac{|\text{actual} - \text{predicted}|}{\text{actual}} \times 100\%$$

Materials

Lead shot, Pb, 35–40 g	Balance, centigram (0.01 g precision)
Silicon lumps, Si, 8–10 g	Beakers, 50-mL, or small cups, 3
Tin shot, Sn, 25–30 g	Forceps or tongs
Water	Graduated cylinder, 25-mL
Paper towels	Marking pencil or pen

Safety Precautions

Lead powder is extremely toxic by inhalation and ingestion; lead fume or dust is a possible carcinogen. Using lead shot should not present a powder or dust hazard. Do not work with lead powder. Silicon is flammable in powder form and is slightly toxic. Do not breathe or handle any fine silicon powder remaining on the bottom of the reagent bottle. Wear chemical splash goggles and chemical-resistant gloves and apron. Wash your hands with soap and water before leaving the laboratory.

There may be some congestion at the balances, especially if fewer than three balances are available for a class of 30 students working in pairs. To prevent delay, consider providing premeasured masses of solids in containers.

Procedure

1. Label three 50-mL beakers or small containers Si (silicon), Sn (tin) and Pb (lead).

2. Obtain approximately 8 g of silicon chunks in the appropriately labeled beaker. Measure the combined mass of the beaker plus solid to the nearest 0.01-g and record the value in the Data Table. (*Note:* This value is the initial mass for sample 1.)

3. Fill a 25-mL graduated cylinder half-full with water. Measure the initial volume of water and record the value to the nearest 0.1-mL in the Data Table.

Teacher Notes

4. Using forceps or tongs, *carefully* add about one-third of the silicon chunks to the graduated cylinder (enough to raise the water level in the cylinder by at least 1.0 mL). Add the solid slowly, so as to avoid splashing or breaking the glass cylinder.
5. Measure and record the new (final) volume of water plus solid in the graduated cylinder.
6. Measure and record the combined mass of the labeled beaker and remaining solid in the Data Table. (*Note:* This value is the final mass for sample 1.)
7. Repeat steps 4–6 *twice* with the remaining amount of solid in the beaker. Record all initial and final mass and volume data in the Data Table. There should be a total of three sets of mass and volume data (samples 1–3).
8. Empty the water from the graduated cylinder and carefully pour the silicon chunks onto a paper towel and allow them to dry. Do not allow any of the solid to go down the drain.
9. Rinse the graduated cylinder with water to make sure that all of the solid has been removed.
10. Obtain approximately 25 g of tin shot in the appropriately labeled beaker. Measure the initial mass of the container plus solid to the nearest 0.01 g and record the value in the Data Table.
11. Repeat steps 3–9 using tin. Record all initial and final mass and volume data in the Data Table.
12. Obtain approximately 35 g of lead shot in the appropriately labeled beaker. Measure the initial mass of the container plus solid to the nearest 0.01 g and record the value in the Data Table.
13. Repeat steps 3–9 using lead. Record all initial and final mass and volume data in the Data Table.
14. Return the correctly labeled solids to your instructor for reuse.

Students may question why the total amount of solid is different for the three metals. The masses of solids have been chosen to give at least two significant figures in the volume measurements. Some students may be delayed trying to add exactly one-third of the metal pieces in step 4. Advise students that this is an approximation and it is not necessary to be exact.

Teacher Notes

Name: ____________________

Class/Lab Period: ____________________

Density Is a Periodic Property

Data Table

Element	Sample	Initial Mass (g)	Final Mass (g)	Mass of Solid (g)	Initial Volume (mL)	Final Volume (mL)	Volume of Solid (mL)
Silicon	1						
	2						
	3						
Tin	1						
	2						
	3						
Lead	1						
	2						
	3						

Post-Lab Calculations and Analysis *(Use a separate sheet of paper to answer these questions.)*

1. *Complete the Data Table:* Calculate both the mass (initial mass – final mass) and volume (final volume – initial volume) of each sample 1–3 for all three elements, silicon, tin, and lead. Record these results in the Data Table.

2. Using the mass and volume data, calculate the density of each sample 1–3 for all three elements. Construct a Results Table to summarize the results. *Note:* The density of a solid is usually reported in units of g/cm^3. Recall that 1 mL = 1 cm^3.

3. Calculate the average value (mean) of the density calculations 1–3 for each element, silicon, tin, and lead. Record all results in the Results Table. Use the range of density values for each element to estimate "plus-or-minus" (±) error for each average (e.g., 7.0 ±0.2 g/cm^3).

4. On a graph, plot the period number of Si, Sn, and Pb on the x-axis versus the average density of each element on the y-axis. Using a ruler or straightedge, draw a "best-fit" straight line through the data points. Use this "best-fit" straight line to predict the density of germanium.

5. Look up the actual density of germanium in a reference source and calculate the percent error between the predicted and actual values (see PreLab Question #3).

Teachers may want to model for students what their Results Table might look like and how to draw a "best-fit" straight line. Remind students that "best-fit" does NOT mean "connect-the-dots."

Teacher Notes

Density Is a Periodic Property

Master Materials List (for a class of 30 students working in pairs)

Lead shot, Pb, 500–600 g*	Balance, centigram (0.01 g precision), 3
Silicon lumps, Si, 120–150 g*	Beakers, 50-mL, or small cups, 45
Tin shot, Sn, 400–500 g*	Forceps or tongs, 15
Water	Graduated cylinder, 25-mL, 15
Paper towels	Hot plate, optional (to dry element samples for recycling)
Marking pencil or pen	

*Metal may be dried and reused from class to class.

Safety Precautions

Lead powder is extremely toxic by inhalation and ingestion; lead fume or dust is a possible carcinogen. Using lead shot should not present a powder or dust hazard. Take all precautions to avoid obtaining lead powder. Silicon is flammable in powder form and is slightly toxic by ingestion. Avoid grinding the silicon chunks when the element is dried or recycled. Do not breathe or handle any fine silicon powder remaining on the bottom of the reagent bottle. Wear chemical splash goggles and chemical-resistant gloves and apron. Consult current Material Safety and Data Sheets for additional safety, handling, and disposal information.

Disposal

Consult your current *Flinn Scientific Catalog/Reference Manual* for general guidelines and specific procedures governing the disposal of laboratory waste. All of the element samples should be thoroughly dried and stored for repeat use. Avoid excessive handling of the silicon lumps. Do not pulverize the material.

Lab Hints

- Prior to lab, review with students the idea of significant figures and remind them to include the proper number of decimal places in their volume readings. Since minor scale divisions on a 25-mL graduated cylinder are marked every 0.2 mL, volume measurements can be estimated to 0.1 mL.

- The sample masses of silicon, tin, and lead have been chosen to allow two significant figures (±0.1 mL) in the volume measurements and to keep the final water volume below the 25-mL capacity of the graduated cylinder. Larger graduated cylinders (100-mL) may also be used, but the sample masses should be increased in order to obtain reliable precision in the volume measurements.

- If the element samples must be reused for several class periods in one day, it will be helpful to provide a way to dry the solids quickly. This can be done by gently heating the solids using a blow dryer or in a pan on a hot plate.

- This experiment can easily be performed using only two elements, silicon and tin, instead of three. In that case, the density of germanium would be calculated as the average of the two elements. The percent error between the predicted and actual densities is greater if only two elements are used (10–15% vs. 5–10%).

Not all graduated cylinders are created equal. For best results, use graduated cylinders that have at least 0.2-mL minor scale divisions. Many teachers may have tin and lead strips that can be used instead of metal shot. Strips are easier to dry and reuse from one class period to another.

Teacher Notes

Teaching Tips

- The water-displacement method for determining the density of solids is discussed in Volume 1, *Introduction to Chemistry,* Flinn ChemTopic® Labs. (See the "Discovering Density" experiment.) Similarly, the concepts of accuracy and precision are the subject of the "Introduction to Measurement" experiment in *Introduction to Chemistry*.
- The "best-fit" straight line should include as many of the data points as possible. This lab provides excellent data to introduce or reinforce the use of graphing calculators and/or graphical analysis programs.
- Challenge students to suggest other properties of the elements, in addition to period number, that can be used to predict the density of germanium within the same group of elements. For example, density can be graphed against either atomic mass or atomic number. Since atomic masses were known at the time of Mendeleev, this reasoning is "fair" and in keeping with how Mendeleev might have predicted the properties of his missing elements. Atomic mass gives a less accurate prediction of the density of germanium than does the period number. Correlation of atomic number with density actually gives the most accurate prediction of the density of germanium. At the time of Mendeleev, of course, there was no concept of atomic number, since protons had not yet been discovered (and would not be for another 40 years or so). Sample graphs of density versus atomic mass and atomic number are shown in the Supplementary Information section.
- Can you predict the density of germanium by looking at the trend across the fourth period of metals, transition metals, and semimetals (K, Ca, Sc, Ti, V . . .) in the periodic table? Consider assigning this as an optional, extra-credit exercise for students to perform outside of class. A reference graph has been included in the Supplementary Information section.

Answers to Pre-Lab Questions *(Students answers will vary.)*

1. One of the elements Mendeleev predicted was *eka*-aluminum, corresponding to a gap in the fourth row or period of the Group IIIA elements, between aluminum and indium. The density of aluminum (period 3) is 2.70, that of indium (period 5) 7.31, and that of thallium (period 6) 11.85 g/cm^3. Make a graph of period number on the x-axis versus density on the y-axis for each of these elements.

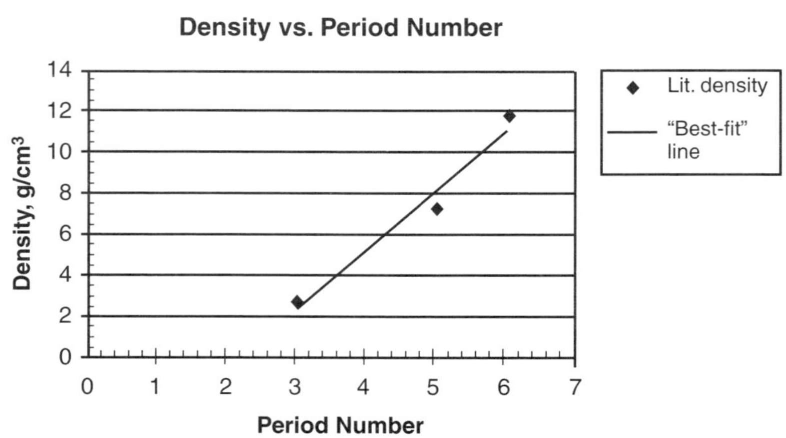

Teacher Notes

2. Use your graph to predict the density of *eka*-aluminum. What known element in the modern Periodic Table corresponds to *eka*-aluminum? Look up the density of the modern element in a reference source and record its actual and predicted density values.

   Using the "best-fit" straight line through the data, the predicted density of the Group III element in period 4 of the Periodic Table would be about 5.3 g/cm^3. Gallium is the modern element corresponding to eka-aluminum, whose existence and properties were predicted by Mendeleev. The actual (literature) density of gallium is 5.90 g/cm^3.

3. How do the actual and predicted density values compare? Use the following equation to calculate the percent error between the predicted and actual values for the density of *eka*-aluminum.

$$\text{percent error} = \frac{|\text{actual} - \text{predicted}|}{\text{actual}} \times 100\%$$

   The predicted density is smaller than the actual value.

$$\textit{percent error} = \frac{|5.90 - 5.3|}{5.90} \times 100\% = 10\%$$

Sample Data

Data Table. *Student data will vary.*

Element	Sample	Initial Mass (g)	Final Mass (g)	Mass of Solid (g)	Initial Volume (mL)	Final Volume (mL)	Volume of Solid (mL)
Silicon	1	37.69	34.97	2.72	14.8	16.0	1.2
	2	34.97	32.04	2.93	16.0	17.4	1.4
	3	32.04	29.59	2.45	17.4	18.5	1.1
Tin	1	57.79	48.28	9.51	15.0	16.3	1.3
	2	48.28	37.85	10.43	16.3	17.8	1.5
	3	37.85	29.72	8.13	17.8	18.9	1.1
Lead	1	72.79	59.94	12.85	15.2	16.4	1.2
	2	59.94	44.33	15.61	16.4	17.8	1.4
	3	44.33	29.05	15.28	17.8	19.2	1.4

Answers to Post-Lab Calculations and Analysis *(Student answers will vary.)*

1. Complete the Data Table: Calculate both the mass (initial mass – final mass) and volume (final volume – initial volume) of each sample 1–3 for all three elements, silicon, tin, and lead. Record these results in the *Data Table*.

   See the Sample Data section.

2. Using the mass and volume data, calculate the density of each sample 1–3 for all three elements. Construct a *Results Table* to summarize the results. *Note:* The density of a solid is usually reported in units of g/cm^3. Recall that 1 mL = 1 cm^3.

Results Table

Element	Density (1) g/cm^3	Density (2) g/cm^3	Density (3) g/cm^3	Average Density g/cm^3
Silicon	*2.3*	*2.1*	*2.2*	*2.2 ±0.1*
Tin	*7.3*	*7.0*	*7.4*	*7.2 ±0.2*
Lead	*10.7*	*11.2*	*10.9*	*10.9 ±0.3*

3. Calculate the average value (mean) of the density calculations 1–3 for each element, silicon, tin, and lead. Record all results in the Results Table. Use the range of density values for each element to estimate "plus-or-minus" (±) error for each average (e.g., 7.0 ±0.2 g/cm^3).

   See Results Table.

4. On a graph, plot the period number of Si, Sn, and Pb on the x-axis versus the average density of each element on the y-axis. Using a ruler or straightedge, draw a "best-fit" straight line through the data points. Use this "best-fit" straight line to predict the density of germanium.

   See the sample graph. Drawing a horizontal line from the period number of germanium (4) across to the y-axis gives a predicted density of germanium equal to 4.9 g/cm^3.

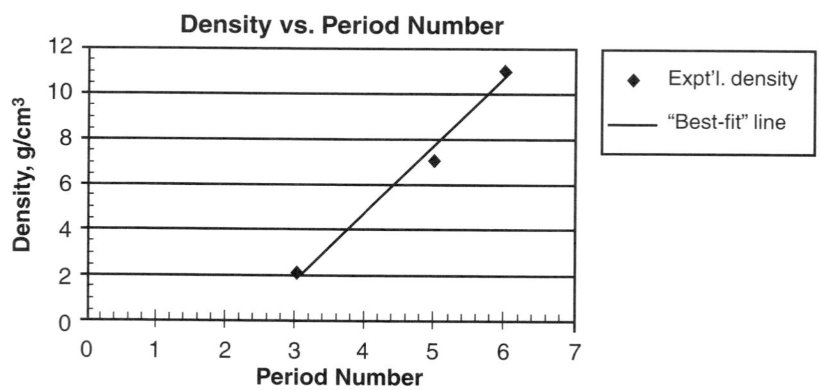

5. Look up the actual density of germanium in a reference source and calculate the percent error between the predicted and actual values (see PreLab Question #3).

   The literature density of germanium (from the CRC Handbook of Chemistry and Physics*) is 5.32 g/cm^3.*

$$\text{percent error} = \frac{|\text{actual} - \text{predicted}|}{\text{actual}} \times 100\% = \frac{5.32 - 4.9}{5.32} \times 100\% = 8\%$$

Teacher Notes

Supplementary Information

Sample Graph. Density vs. Period Number for Group IVA Elements

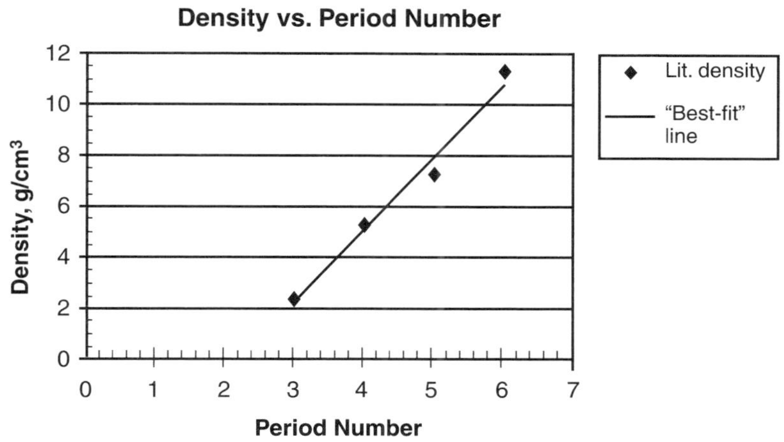

Sample Graph. Density vs. Atomic Mass for Group IVA Elements

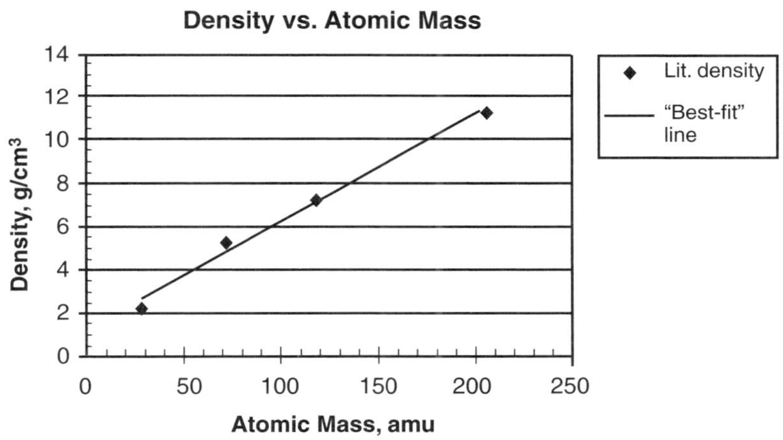

Sample Graph. Density vs. Atomic Number for Group IVA Elements

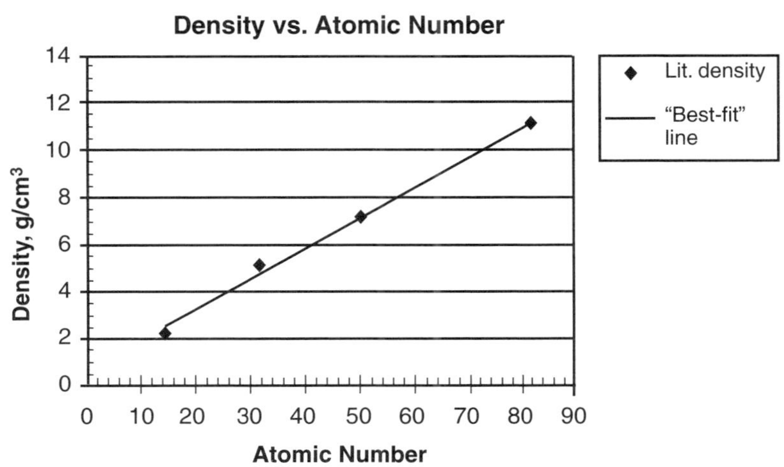

Sample Graph. Density vs. Atomic Mass for the Period 4 Metals, Transition Metals, and Semimetals

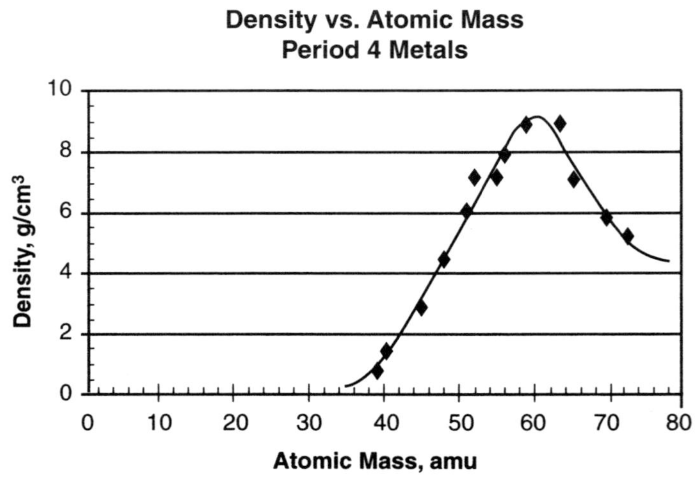

Teacher Notes

Periodic Trends and the Properties of Elements
The Alkaline Earth Metals

Introduction

The periodic table is the most recognized symbol of chemistry across the world. It is a valuable tool that allows scientists not only to classify the elements but also to explain and predict their properties. Similarities and differences among the elements give rise to so-called periodic trends, both across rows and within columns of the periodic table. Recognizing periodic trends in the physical and chemical properties of the elements is key to understanding the full value of the periodic table.

Concepts

- Periodic table
- Group vs. period
- Activity series
- Periodic trends
- Alkaline earth metals
- Double replacement reaction

Background

The modern periodic table lists more than 112 elements, of which 92 are naturally occurring. Of these 92 elements, the eight most abundant elements together account for more than 98 percent of the mass of the Earth's crust, oceans, and atmosphere. Two of the eight most abundant elements on Earth are calcium and magnesium, which are present in both mountains and minerals, seawater and seashells. Calcium and magnesium are members of the Group IIA family of elements, the alkaline earth metals. Elements that share similar properties are arranged together within vertical columns, called groups or families, in the periodic table.

The *alkaline earth metals*—beryllium, magnesium, calcium, strontium, barium, and radium—are a reactive group of metals. Because they combine easily with many other elements, the alkaline earth elements are not found on Earth in the form of their free metals. They exist in nature in the form of ionic compounds, such as calcium carbonate, $CaCO_3$. Calcium carbonate occurs naturally in limestone, marble, as well as seashells.

The alkaline earth metals react with water, acids and bases, and many nonmetals, including oxygen, sulfur, and the halogens. The ease with which a metal reacts is called the activity of the metal. By comparing how fast or how vigorously different metals react, it is possible to rank the metals in order from most active to least active. This ranking—called the *activity series* of the metals—shows clear periodic trends, both within a group and across a period of elements in the periodic table.

Periodic trends are also observed in the solubility of alkaline earth metal compounds. Although their compounds with halide anions are all water-soluble, alkaline earth metal compounds with other anions do not always dissolve in water. The solubility of alkaline earth metal compounds with different anions can be tested by carrying out so-called *double replacement reactions*. Reaction of calcium chloride with sodium carbonate, for example, leads to an exchange of anions between the two metals to give calcium carbonate, which is insoluble in water and precipitates out as a solid when the two solutions are mixed. The chemical equation for this reaction is shown in *Equation 1,* where the abbreviations (aq) and (s) refer to aqueous solutions and solid precipitates, respectively.

$CaCl_2(aq)$	+	$Na_2CO_3(aq)$	→	$CaCO_3(s)$	+	$2NaCl(aq)$	*Equation 1*
calcium chloride		*sodium carbonate*		*calcium carbonate*		*sodium chloride*	

Experiment Overview

The purpose of this experiment is to identify periodic trends in the activity and solubility of the alkaline earth metals. In Part A, the reactions of magnesium, calcium, and aluminum with water and acids will be compared in order to determine the trend in metal activity within a group (Mg vs. Ca) and across a period (Mg vs. Al) in the periodic table. In Part B, the solubility of magnesium, calcium, strontium, and barium compounds will be studied and used to identify an unknown alkaline earth metal.

Pre-Lab Questions

1. Read the entire procedure and the recommended safety precautions. Do you think extra pieces of calcium or magnesium metal should be disposed of down the drain? Why or why not?

2. The ionization energy of an element is defined as the amount of energy required to remove an electron from an individual atom. The following table gives the ionization energy (in units of kilojoules per mole) for five metals, listed in alphabetical order. Locate each of these metals on the periodic table and arrange them in order of rows and columns as in the periodic table.

   (a) Describe the periodic trend in the ionization energy of elements within a group.

   (b) Describe the periodic trend in the ionization energy of elements across a period.

Metal	Calcium	Magnesium	Potassium	Sodium	Strontium
Ionization Energy	590	738	419	496	549

Materials

Aluminum foil, Al, 2-cm square, 2
Barium chloride, $BaCl_2$, 0.1 M, 3 mL
Calcium, turnings, Ca, 2 pieces
Calcium chloride, $CaCl_2$, 0.1 M, 3 mL
Magnesium ribbon, Mg, 1-cm piece, 2
Magnesium chloride, $MgCl_2$, 0.1 M, 3 mL
Strontium chloride, $SrCl_2$, 0.1 M, 3 mL
Unknown metal chloride solution, 0.1 M, 3 mL
Water, distilled or deionized
Matches (optional)
Hydrochloric acid, HCl, 0.5 M, 3 mL
Potassiun iodate, KIO_3, 0.2 M, 5 mL
Sodium carbonate, Na_2CO_3, 1 M, 5 mL
Sodium sulfate, Na_2SO_4, 1 M, 5 mL
Forceps
Litmus paper, red, 3 pieces
Paper, white and black, 1 sheet each
Pipets, Beral-type, 10
Reaction plate, 24-well
Thermometer

Beral-type pipets are disposable polyethylene transfer pipets. The graduated or thin-stem versions are excellent for this activity. Consider setting up several stations to dispense the reagents for both Parts A and B. If the reagents are placed in dropper bottles, the transfer pipets will not be needed.

Teacher Notes

Safety Precautions

Calcium and magnesium are reactive, flammable solids and possible skin irritants. Use forceps or a spatula to handle these metals. Hydrochloric acid is toxic by ingestion and inhalation and is corrosive to skin and eyes; avoid contact with body tissues. Strontium and barium compounds are toxic by ingestion. Potassium iodate solution is moderately toxic and a strong irritant. Avoid contact of all chemicals with eyes and skin. Wear chemical splash goggles and chemical-resistant gloves and apron. Always wash hands thoroughly before leaving the laboratory.

Procedure

Part A. Activity of Metals

1. In a weighing dish or small beaker, obtain 2 small pieces of calcium turnings.
2. Obtain 2 small pieces of magnesium ribbon, approximately 1-cm each, and a short piece of aluminum foil.
3. Place a 24-well reaction plate on top of a sheet of white paper, as shown in the following figure. Note that each well is identified by a unique combination of a letter and a number, where the letter refers to a horizontal row and the number to a vertical column.

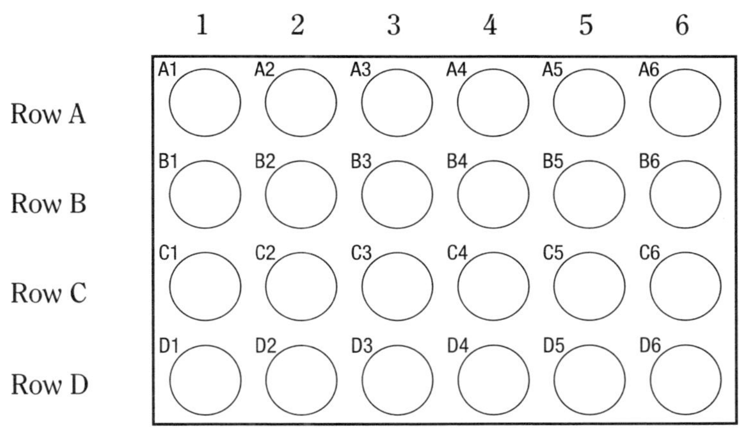

4. Use a pipet to add 20 drops of distilled water to wells A1–A3.
5. Test the water in wells A1–A3 with a piece of red litmus paper and record the initial color for this "litmus test" in Data Table A.
6. Use forceps to add one piece of calcium (step 1) to well A1.
7. Use forceps to add one piece of magnesium ribbon to well A2.
8. Tear off a 2-cm piece of aluminum foil and roll it into a loose ball. Add the aluminum metal to well A3.
9. Observe each well and record all immediate observations in Data Table A. If no changes are observed in a particular well, write NR (No Reaction) in the data table.
10. Test the water in wells A1–A3 with a piece of red litmus paper and record the color changes for this litmus test in Data Table A.

11. Continue to watch each well for 1–2 minutes. Record any additional observations comparing the *rates of reaction* in Data Table A.
12. Use a pipet to add 20 drops of 0.5 M HCl to wells C1–C3 (the first three wells in Row C). Measure the initial temperature of the solutions in well C1–C3 and record the values as an "observation" in Data Table A.
13. Use forceps to add one piece of calcium turnings (Step 1) to well C1.
14. Use forceps to add one piece of magnesium ribbon to well C2.
15. Tear off a 2-cm piece of aluminum foil and roll it into a loose ball. Add the aluminum metal to well C3.
16. Observe each well and record all immediate observations in Data Table A. If no changes are observed in a particular well, write NR in the data table.
17. Using a thermometer, measure the temperature of each solution in wells C1–C3. Record the temperature of each solution as an observation in Data Table A.
18. *(Optional)* Is there evidence that a gas is being produced in wells C1–C3? Test the combustion property of the gas by bringing a lit match to the space just above each well C1–C3. Record any observations for this "match test" in Data Table A.
19. Continue to watch each well for 1–2 minutes. Record any additional observations comparing the *rates of reaction* in Data Table A.
20. Dispose of the well contents as instructed by your teacher. Rinse the reaction plate with distilled water before using the plate again in Part B.

Part B. Solubility of Alkaline Earth Metal Compounds

21. Place the 24-well reaction plate on top of a sheet of *black paper*.
22. Referring to Data Table B as a guide, use a pipet to add 20 drops of alkaline earth metal solutions to the appropriate wells, as follows:
    - Magnesium chloride to wells A1–C1
    - Calcium chloride to wells A2–C2
    - Strontium chloride to wells A3–C3
    - Barium chloride to wells A4–C4.
23. Use a clean pipet to add 20 drops of the unknown alkaline earth metal solution to wells A5–C5.
24. Referring to Data Table B as a guide, use a clean pipet to add 20 drops of testing solution to the appropriate wells, as follows:
    - Sodium carbonate to wells A1–A5
    - Sodium sulfate to wells B1–B5
    - Potassium iodate to wells C1–C5
25. Record observations in Data Table B as follows: if a solid forms in a well, write PPT (precipitate) in the appropriate circle in the data table. If no solid is observed, write NR (no reaction) in the appropriate circle in the data table.
26. Dispose of the contents of the reaction plate as instructed by your teacher.

Teacher Notes

See the Disposal section for disposal recommendations. Because unreacted metal pieces remain in the wells after Part A, the reaction plate contents should be poured into a beaker containing dilute HCl. The reaction plate contents from Part B can be washed down the drain, or the contents can be poured onto several layers of paper towels in a large tray. When the paper towels are dry, they are suitable for disposal in a landfill according to Flinn Suggested Disposal Method #26a.

Teacher Notes

Name: ______________________________

Class/Lab Period: ______________________________

Periodic Trends and the Properties of Elements

Data Table A. *Activity of Metals*

		Calcium	Magnesium	Aluminum
Reaction with H_2O	Observations			
	Litmus test			
Reaction with HCl	Observations			
	Match test (optional)			

Data Table B. *Solubility of Alkaline Earth Metal Compounds*

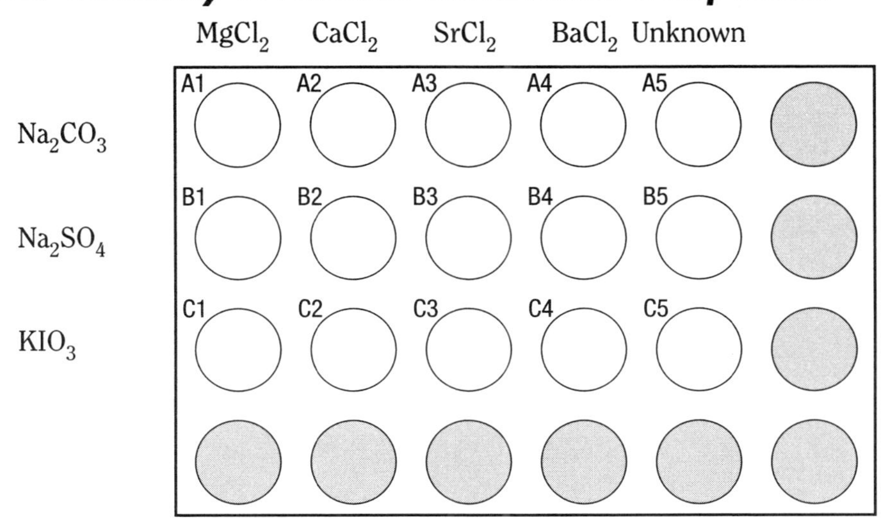

Teacher Notes

Post-Lab Questions *(Use a separate sheet of paper to answer the following questions.)*

1. Which group IIA metal, magnesium or calcium, is more active? Cite your evidence.
2. Which period 3 metal, magnesium or aluminum, is more active? Cite your evidence.
3. Rank the three metals tested in Part A from most active to least active.
4. Write a general statement that describes the periodic trend in metal activity across a group (vertical column) of the periodic table.
5. Write a general statement that describes the periodic trend in metal activity within a period (horizontal row) of the periodic table.
6. Locate the following metals on the periodic table: *magnesium, potassium,* and *sodium*. Based on your answers to Questions #4 and 5, rank these metals in order of their expected activity, from most active to least active.
7. Litmus paper changes color in acidic (red) and basic (blue) solutions. The word alkaline is a synonym for basic. Why are the two words "alkaline" and "earth" used to name the Group IIA metals?
8. In Part B, which alkaline earth metal formed the most precipitates? The fewest?
9. Write a general statement that describes the periodic trend in the solubility of alkaline earth metal compounds.
10. Use the solubility pattern observed for the known and unknown alkaline earth compounds in Part B to deduce the identity of the unknown alkaline earth metal. Explain your reasoning.
11. *(Optional)* Using Equation 1 in the Background Section as an example, write a chemical equation for each precipitate-forming reaction that was observed for *strontium* in Part B. Include the abbreviations (aq) and (s) to show what compound is responsible for the precipitate in each case.

Teacher Notes

Periodic Trends and the Properties of Matter

Microscale

Master Materials List *(for class of 30 students working in pairs)*

Aluminum foil, Al, 1 sheet,* 30 × 30 cm
Barium chloride, $BaCl_2$, 0.1 M, 45 mL
Calcium, turnings, Ca, 8 g
Calcium chloride, $CaCl_2$, 0.1 M, 45 mL
Magnesium ribbon, Mg, 30 cm
Magnesium chloride, $MgCl_2$, 0.1 M, 45 mL
Strontium chloride, $SrCl_2$, 0.1 M, 45 mL
Unknown metal chloride solution, 0.1 M, 45 mL
Water, distilled or deionized, 75 mL
Matches (optional)
Hydrochloric acid, HCl, 0.5 M, 45 mL
Potassium iodate, KIO_3, 0.2 M, 75 mL
Sodium carbonate, Na_2CO_3, 1 M, 75 mL
Sodium sulfate, Na_2SO_4, 1 M, 75 mL
Forceps, 15
Litmus paper, red, 1 vial
Pipets, Beral-type, 150
Reaction plate, 24-well, 15
Thermometer, 15

*Household-type aluminum foil is fine for this application.

Preparation of Solutions *(for class of 30 students working in pairs)*

Barium chloride, 0.1 M: Dissolve 2.4 g of barium chloride dihydrate ($BaCl_2{\cdot}2H_2O$) in about 60 mL of distilled or deionized water. Stir to dissolve and dilute to 100 mL with water.

Calcium chloride, 0.1 M: Dissolve 1.5 g of calcium chloride dihydrate ($CaCl_2{\cdot}2H_2O$) in about 60 mL of distilled or deionized water. Stir to dissolve and dilute to 100 mL with water.

Magnesium chloride, 0.1 M: Dissolve 2.0 g of magnesium chloride hexahydrate ($MgCl_2{\cdot}6H_2O$) in about 60 mL of distilled or deionized water. Stir to dissolve and dilute to 100 mL with water.

Strontium chloride, 0.1 M: Dissolve 2.7 g of strontium chloride hexahydrate ($SrCl_2{\cdot}6H_2O$) in about 60 mL of distilled or deionized water. Stir to dissolve and dilute to 100 mL with water.

Potassium iodate, 0.2 M: Dissolve 4.3 g of potassium iodate (KIO_3) in about 60 mL of distilled or deionized water. Stir to dissolve and dilute to 100 mL with water.

Sodium sulfate, 1 M: Dissolve 32 g of sodium sulfate decahydrate ($Na_2SO_4{\cdot}10H_2O$) in about 60 mL of distilled or deionized water. Stir to dissolve and dilute to 100 mL with water.

Sodium carbonate, 1 M: Dissolve 12.4 g of sodium carbonate monohydrate ($Na_2CO_3{\cdot}H_2O$) in about 60 mL of distilled or deionized water. Stir to dissolve and dilute to 100 mL with water.

Most of these solutions are available in ready-made form from Flinn Scientific. See the Master Materials Guide in the Supplementary Information section of this lab manual.

Safety Precautions

Calcium and magnesium are reactive, flammable solids and are possible skin irritants. Calcium may react violently with water and acids. Dispense only small sample sizes of these metals and use forceps or a spatula to handle them. Hydrochloric acid is toxic by ingestion and inhalation and is corrosive to skin and eyes; avoid contact with body tissues. Strontium and barium compounds are toxic by ingestion. Potassium iodate solution is moderately toxic and a strong irritant. Avoid contact of all chemicals with eyes and skin. Wear chemical splash goggles and chemical-resistant gloves and apron. Consult current Material Safety Data Sheets for additional safety, handling, and disposal information.

Disposal

Consult your current *Flinn Scientific Catalog/Reference Manual* for general guidelines and specific procedures governing the disposal of laboratory waste. There will be unreacted magnesium metal remaining in the reaction plates, even in the HCl well. Do NOT allow students to rinse the contents of their reaction plates in Part A down the drain. Students should use a Beral-type pipet to remove and rinse the contents of the reaction plate into a central "Metal Waste" beaker containing 1 M hydrochloric acid. Dispose of this "Metal Waste" solution down the drain with plenty of excess water according to Flinn Suggested Disposal Method #26b only after you are sure all of the metal has reacted.

The reaction plate contents from Part B can be washed down the drain with excess water according to Flinn Suggested Disposal Method #26b. Excess barium chloride solution can be disposed of by converting it to insoluble barium sulfate according to Flinn Suggested Disposal Method #27h. Excess potassium iodate can be disposed of by reduction according to Flinn Suggested Disposal Method #12a.

Lab Hints

- Dispense only small pieces of calcium turnings and magnesium ribbon. Ideally, one small piece of calcium turnings should have a mass of about 0.05 g. Magnesium turnings may be substituted for ribbon, but do NOT use magnesium powder. Many teachers prefer to dispense these active metals from the front of the room so that they can more easily monitor the students.

- The concentration of hydrochloric acid in Part A allows students to observe vigorous but still safe reactions with both calcium and magnesium. Do NOT increase the concentration of acid.

- Aluminum does not react with 0.5 M HCl. In order to demonstrate a positive reaction with aluminum, an acid concentration greater than about 3 M would be needed. If desired, the reaction of aluminum metal with more concentrated hydrochloric acid may be demonstrated by the teacher to show that aluminum also exhibits typical metal activity when the acid concentration is high enough. Do NOT allow students to test the higher HCl concentration. Reaction of calcium with more concentrated acid is violent, unsafe, and potentially explosive.

Teacher Notes

- Review with students the nature of the optional burning match test to identify hydrogen. The test results are not needed to answer the Post-Lab questions or to achieve the experiment objectives.

- In order to avoid exposing students to more hazards, only the minimum number of solubility tests has been included in Part B. Other typical solubility tests for alkaline earth metal cations involve using sodium oxalate, ammonium hydroxide, and ammonium chromate to precipitate oxalate, hydroxide and chromate salts, respectively. Sodium oxalate gives the same precipitate pattern with alkaline earth metals as does sodium carbonate. Similarly, ammonium chromate gives the same solubility pattern as potassium iodate. Chromate compounds are known carcinogens and are not recommended for use in this experiment. Ammonium hydroxide can be used to show a positive result for magnesium ion (in the current procedure, magnesium is identified by negative test results with all three testing solutions). Hydroxides are the exception to the rule in the solubility pattern of alkaline earth metals—magnesium hydroxide and barium hydroxide form precipitates, while calcium and strontium hydroxide do not.

- For the reasons cited above, magnesium chloride does not make a good unknown in Part B. Prepare three unknowns A, B, and C—containing calcium, strontium, and barium chloride, respectively—and assign them at random to students.

- The solubility tests can also be done on a sheet of acetate transparency. On a sheet of white paper, draw large circles as in Data Table B to represent the reactions of each of the alkaline earth metal compounds with the three testing solutions. Place an acetate transparency sheet on top of this master data table sheet, then add a few drops of each reagent to the appropriate circle and observe. The transparency sheet can be rinsed and reused.

Teaching Tips

- The following assumptions have been made in designing the experiment objectives and writing appropriate post-Lab questions. (a) Students should be able to recognize charges on common ions and write correct formulas for ionic compounds. (b) Students may not be familiar with the classification of chemical reactions and may not have experience writing complete or net ionic equations. Teachers who have already covered these topics may wish to assign additional post-lab questions to build on prior learning.

- This lab provides a good opportunity to tell the story of Marie Curie and her discovery of radium. It is a heroic story of dedication, patience, and persistence. Starting with *one ton* of a special uranium-depleted ore, Marie Curie carried out a series of exhausting large-scale precipitation reactions. She obtained *100 milligrams* of a new, radioactive element—radium, the heaviest member of the alkaline earth metals. The separation worked because of the periodic trend identified in Part B of this experiment—the solubility of alkaline earth metals decreases as you go down the column in the Periodic Table. Radium chloride is *ever-so-slightly* less soluble than barium chloride. Working with concentrated hydrochloric acid, Marie Curie carried out many, many repeat *fractional crystallization* reactions with a mixture of barium and radium chlorides from the ore. With each successive crystallization, the precipitate gradually became enriched in the amount of the less soluble radium fraction, until finally almost pure radium chloride was obtained.

Answers to Pre-Lab Questions *(Student answers will vary.)*

1. Read the entire procedure and the recommended safety precautions. Do you think extra pieces of calcium or magnesium metal should be disposed of down the drain? Why or why not?

   No! Solids that do not dissolve in water should never be disposed of down the drain—they will clog the plumbing. Note to teachers: These metals may react violently with water. Disposing of these metals down the drain may start fires or even explosions.

2. The ionization energy of an element is defined as the amount of energy required to remove an electron from an atom. The following table gives the ionization (in units of kilojoules per mole) for five metals, listed in alphabetical order. Locate each of these metals on the periodic table and arrange them in the same order of rows and columns as in the periodic table.

   (a) Describe the periodic trend in the ionization energy of elements within a group.

   (b) Describe the periodic trend in the ionization energy of elements within a period.

Metal	Calcium	Magnesium	Potassium	Sodium	Strontium
Ionization Energy	590	738	419	496	549

	Group IA	Group IIA
Period 2	*Na*	*Mg*
Period 3	*K*	*Ca*
Period 4		*Sr*

(a) Ionization energy decreases as you go down a group in the periodic table.

(Mg>Ca>Sr).

(b) Ionization energy increases from left to right across a period in the periodic table.

(Na<Mg and K<Ca)

If you want to add aluminum to the metals being compared in Question #2, remember that the ionization energy of Al is actually less than that of Mg. This "inversion" occurs because the 2p orbital is higher in energy than the 2s orbital.

Teacher Notes

Sample Data

Student data will vary.

Data Table A. *Activity of Metals*

		Calcium	Magnesium	Aluminum
Reaction with H_2O	Observations	Solution bubbles, fizzes, foams, and smokes. Metal is gone, replaced by crumbly white solid. Final mixture is cloudy, milky white.	Some slight bubbling seen on surface of Mg metal.	NR
	Litmus test	Changes from red (before reaction) to blue (after reaction).	No change (red)	No change (red)
Reaction with HCl	Observations	Solution bubbles, fizzes, foams, smokes and boils! Almost out-of-control. Temp increases from 20 °C to 80 °C. Metal disappears; final mixture is chalky white.	Solution bubbles, fizzes. Gas given off. Temperature increases from 20 °C to 35 °C.	NR
	Match test (optional)	Large "pop" and flame goes out.	NR	NR

Data Table B. Solubility of Alkaline Earth Compounds

	$MgCl_2$	$CaCl_2$	$SrCl_2$	$BaCl_2$	Unknown
Na_2CO_3	A1 NR	A2 PPT	A3 PPT	A4 PPT	A5 PPT
Na_2SO_4	B1 NR	B2 NR	B3 PPT	B4 PPT	B5 PPT
KIO_3	C1 NR	C2 NR	C3 NR	C4 PPT	C5 NR

Answers to Post-Lab Questions

1. Which group IIA metal, magnesium or calcium, is more active? Cite your evidence.

   Ca is more reactive than Mg. Ca reacted with both water and HCl (sometimes violently in the case of HCl), whereas Mg reacted only with HCl, and then its reaction was much less vigorous than in the case of Ca.

2. Which period 3 metal, magnesium or aluminum, is more active? Cite your evidence.

   Mg is more reactive than Al. Mg reacted noticeably with HCl, while Al exhibited little or no evidence of reaction (a few bubbles here and there).

3. Rank the three metals tested in Part A from most active to least active.

   Ca>Mg>Al

4. Write a general statement that describes the periodic trend in metal activity within a group (vertical column) of the periodic table.

   Metal activity increases as you go down a group in the periodic table, i.e., as the atomic weight of the element increases.

5. Write a general statement that describes the periodic trend in metal activity within a period (horizontal row) of the periodic table.

   Metal activity decreases from left to right across a period in the periodic table.

6. Locate the following metals on the periodic table: *magnesium, potassium,* and *sodium*. Based on your answers to Questions #4 and 5, rank these metals in order of their expected activity, from most active to least active.

   K>Na>Mg

7. Litmus paper changes color in acidic (red) and basic (blue) solutions. The word alkaline is a synonym for basic. Give two reasons why the Group IIA metals are called *alkaline earth* metals.

   Litmus paper changed from red to blue (basic) when it was dipped into the water solution after reaction with calcium metal. This indicates that calcium reacted with water to form an alkaline solution. The name alkaline earth metal thus reflects two facts: they produce alkaline solutions and they are important components of the Earth's crust and oceans.

8. In Part B, which alkaline earth metal formed the most precipitates? The fewest?

   Barium formed precipitates with all of the testing solutions, while magnesium did not form a precipitate with any of the solutions tested.

The Ultimate Element Crossword Puzzle

At the bottom of this page is an alphabetical listing of the first 103 elements. Sixty-two of them have been made into a crossword puzzle below. Use the clues provided to fill in the appropriate letters. The clues themselves contain over 70 word plays. Circle all the ones you find. For example, in one of the clues the words "our scenic" are used and refer to the answer to that clue: "arsenic."

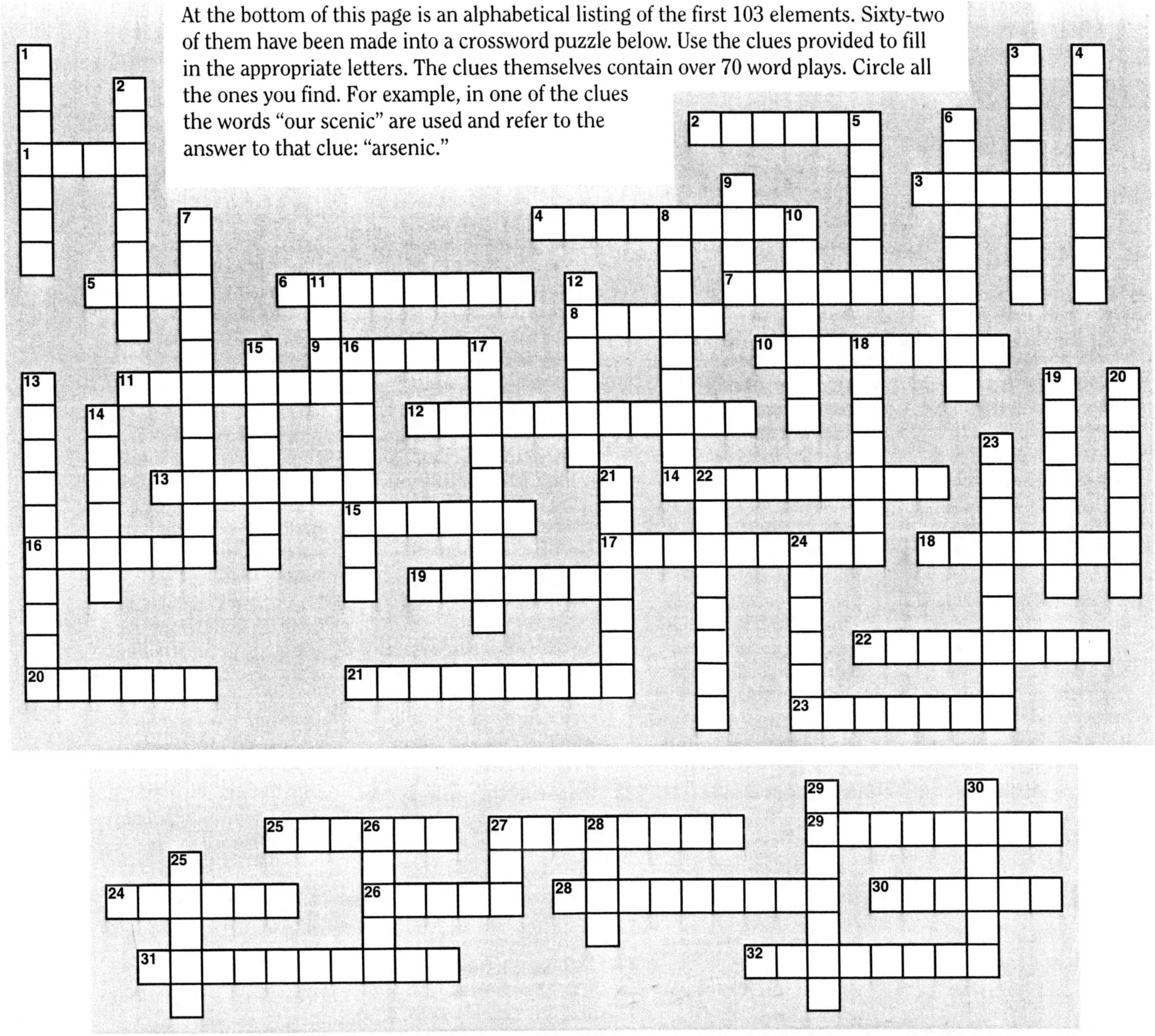

Teacher Notes

9. Write a general statement that describes the periodic trend in the solubility of alkaline earth metal compounds.

   The solubility of alkaline earth metal compounds decreases as you go down the column in the periodic table, i.e., the solubility of their compounds decreases as the atomic weight of an alkaline earth metal increases.

10. Use the solubility pattern observed for the known and unknown alkaline earth compounds in Part B to deduce the identity of the unknown alkaline earth ion. Explain your reasoning.

    Answers will vary. The unknown metal described in the Sample Data is strontium. It showed the same solubility pattern as strontium chloride with each of the three testing solutions. Thus, strontium iodate is soluble, while strontium carbonate and strontium sulfate are not.

11. *(Optional)* Using Equation 1 in the Background Section as an example, write a chemical equation for each precipitate-forming reaction that was observed for *strontium* in Part B. Include the abbreviations (aq) and (s) to show what compound is responsible for the precipitate in each case.

    $$SrCl_2(aq) + Na_2CO_3(aq) \rightarrow SrCO_3(s) + 2NaCl(aq)$$

    $$SrCl_2(aq) + Na_2SO_4(aq) \rightarrow SrSO_4(s) + 2NaCl(aq)$$

Teacher Notes

Teacher Notes

All in the Family
The Halogens and Their Compounds

Introduction

In every family there are both similarities and differences. The same is true in chemical families as well. Elements that share similar chemical properties are arranged in vertical columns, called groups or families, in the modern periodic table. The Group VIIA elements, for example, consisting of fluorine, chlorine, bromine, iodine, and astatine, are called the halogens. What are the similarities and differences in the chemical properties of the halogens and their compounds?

Concepts

- Periodic table
- Halogens
- Activity series
- Groups or families
- Periodic trends
- Single replacement reactions

Background

The *halogens* are a reactive group of nonmetals. The first two members of the group, fluorine and chlorine, are gases at room temperature. As the atomic mass of the halogens increases, the physical state of the elements also changes—bromine is a liquid at room temperature, while iodine is a solid. This change from gas to liquid to solid as the atomic mass of the halogens increases is easily observed. There are other "periodic" trends in the properties of the halogens that are not as easily observed or measured. Thus, it is known that the atomic size (radius) of the halogens increases going down the column of elements, but that the ionization energy decreases in the same direction. *Periodic trends* in the physical properties of the halogens are mirrored by trends in their chemical properties as well.

The halogens exist in nature mainly in the form of ionic compounds containing halide anions. Calcium fluoride, for example, occurs in mineral deposits, while sodium chloride, sodium bromide, and sodium iodide occur naturally in seawater. (Astatine is a radioactive, unstable, and very rare element.) The halogens readily combine with many other elements, including metals, hydrogen, and oxygen. Indeed, the halogens do not exist in nature as free elements. The halogens are prepared for industrial use by passing an electric current through sodium halide solutions. When the halogens are prepared in this way, they are found to exist as diatomic molecules—F_2, Cl_2, Br_2, and I_2. The free halogens are generally quite toxic and must be handled with caution.

Students frequently confuse the –ine and –ide endings for the halogens and their halide ions, respectively. It may be helpful to review this terminology during the pre-lab discussion.

One of the many types of reactions that the halogens (X_2) undergo are so-called single replacement reactions with metal halides (MY), as shown in *Equation 1*.

$$X_2 + 2MY \rightarrow 2MX + Y_2 \qquad \textit{Equation 1}$$

These reactions will take place in one direction only. The reaction of one halogen (X_2) with a different metal halide (MY) will occur only if X_2 is more reactive than Y_2. If X_2 is less reactive than Y_2, the reaction shown in Equation 1 will not take place. The *activity series* of the halogens lists the halogens in order from the most reactive to the least reactive. By comparing the pair-wise reactions of chlorine, bromine, and iodine with sodium chloride, sodium bromide, and sodium iodide, it should be possible to determine the periodic trend in the chemical reactivity of the halogens.

Teacher Notes

Experiment Overview

The purpose of this experiment is to explore the similarities and differences in the chemical properties of the halogens. In Parts A and B, the reactions of chlorine, bromine, and iodine with sodium chloride, sodium bromide, and sodium iodide will be studied in order to determine the activity series of the halogens. In Part C, the solubility of metal chlorides, bromides, and iodides will be examined.

Pre-Lab Questions

1. Read the entire procedure and the recommended safety precautions. What hazards are associated with the use of halogens in the lab? How can these hazards be avoided?

2. Chlorine is used in water treatment plants and in swimming pools. Why is chlorine added to drinking water and to swimming pool water? Based on your answer to Question #1, suggest a possible disadvantage to the use of chlorine in water treatment.

3. The following table shows the boiling points of the halogens. Describe in words the periodic trend that is observed.

Halogen	F_2	Cl_2	Br_2	I_2
Boiling point	–188 °C	–34 °C	58 °C	184 °C

Materials

Ammonia water, NH_3, 1.0 M, 6 mL
Bromine water, Br_2 in H_2O, 1 mL
Chlorine water, Cl_2 in H_2O, 1 mL
Iodine solution, I_2, 0.05 M, 1 mL
Hexane, C_6H_{14}, 6 mL
Silver nitrate solution, $AgNO_3$, 0.1 M, 3 mL
Sodium bromide solution, NaBr, 0.1 M, 2 mL
Sodium chloride solution, NaCl, 0.1 M, 2 mL
Sodium iodide solution, NaI, 0.1 M, 2 mL
Marking pencil or pen
Pipets, Beral-type (or eyedroppers), 9
Test tubes, 12 × 75 mm, 6
Test tube rack
Stirring rod (optional)
Stoppers for test tubes, 6
Water, distilled or deionized

Most reference sources list the normal boiling point of iodine as 184 °C. Iodine is well-known to sublime below its melting point at atmospheric pressure. Above its melting point of 113 °C, iodine first melts and then boils at 184 °C.

Safety Precautions

Bromine water and chlorine water are toxic by ingestion and inhalation. Do not breathe bromine and chlorine vapors. Bromine and iodine water are skin and eye irritants. Avoid contact of all chemicals with eyes and skin. Hexane is a flammable organic liquid. Do not allow hexane to come in contact with heat, flames or other sources of ignition. Wear chemical splash goggles and chemical-resistant gloves and apron.

Teacher Notes

Students are often stumped when no reaction is observed. They may lack the experience or confidence to state that no change has been observed. It may be helpful to keep a control set of test tubes from Part A on display in the hood. Sometimes a leading question may be helpful—what color would you expect if a reaction did occur?

Procedure

Part A. Color and Appearance of the Halogens

Hexane is an organic solvent that does not dissolve in water. Chlorine, bromine, and iodine will be identified by their color and appearance in both water and hexane.

1. Place six clean test tubes in a test tube rack. Label the tubes 1–6.
2. Use a pipet to add 10 drops of chlorine water (Cl_2) to tubes 1 and 2.
3. Use a different pipet to add 10 drops of bromine water (Br_2) to tubes 3 and 4.
4. Use a third pipet to add 10 drops of iodine solution (I_2) to tubes 5 and 6.
5. Record the color of each aqueous halogen solution in Data Table A.
6. Add approximately 20 drops of hexane to each tube 1–6.
7. Place a stopper on each tube. Hold the stopper in place with a finger and carefully invert each tube several times to mix the liquids.
8. Use Data Table A to record your observations of each halogen mixture after hexane has been added. (How many layers are present? Which layer is which? What color(s) are they?)

Part B. Reactions of Halogens with Sodium Halides

9. Add 10 drops of sodium bromide (NaBr) solution to the chlorine solution in tube 1. Replace the stopper on the tube and carefully invert the tube several times to mix the liquids. Observe any reaction that occurs—look for changes in color to either layer. Record all observations in Data Table B. If no change is apparent, write NR (no reaction) in the data table.
10. Continue to add 10 drops of the appropriate sodium halide solution to each test tube, as shown in the following chart. Make and record detailed observations as you add each solution and mix the contents of each tube. Complete Data Table B.

Test Tube	**1**	**2**	**3**	**4**	**5**	**6**
Halogen	Cl_2	Cl_2	Br_2	Br_2	I_2	I_2
Sodium Halide	NaBr	NaI	NaCl	NaI	NaCl	NaBr

11. Dispose of the solutions as instructed by your teacher. Rinse the tubes with water before using them again in Part C.

Teacher Notes

Part C. Solubility of Metal Halides

12. Label three small test tubes 1–3.

13. Using a separate pipet for each solution, add 20 drops of 0.1 M sodium halide solution to the appropriate test tube, as shown below.

Test Tube	**1**	**2**	**3**
Sodium Halide	NaCl	NaBr	NaI

14. Use a labeled pipet to add 20 drops of 0.1 M silver nitrate solution ($AgNO_3$) to each test tube, observing carefully as each drop is added. Record observations in Data Table C.

15. Use a labeled pipet to add 30 drops of ammonia water into each test tube, observing carefully as each drop is added. Gently swirl or stir the test tubes to mix the contents. Record observations in Data Table C. Allow any precipitate to settle for 1–2 minutes and record any additional observations.

16. Dispose of the solutions as instructed by your teacher.

Teacher Notes

Name: ______________________________

Class/Lab Period: ______________________________

All in the Family

Data Table A. *Color and Appearance of the Halogens*

Tubes	Halogen	Initial observations	Observations with hexane
1, 2	Cl_2		
3, 4	Br_2		
5, 6	I_2		

Data Table B. *Reactions of Halogens with Sodium Halides*

Tube	Halogen	Sodium Halide	Observations
1	Cl_2	NaBr	
2	Cl_2	NaI	
3	Br_2	NaCl	
4	Br_2	NaI	
5	I_2	NaCl	
6	I_2	NaBr	

Data Table C. *Solubility of Metal Halides*

Tube	Sodium Halide	Observations with $AgNO_3$	Observations with aqueous ammonia
1	NaCl		
2	NaBr		
3	NaI		

Teacher Notes

Post-Lab Questions *(Use a separate sheet of paper to answer the following questions.)*

1. How can each of the halogens—Cl_2, Br_2, and I_2—be detected in the aqueous layer and in the organic (hexane) layer? Based on your observations, are the halogens more soluble in water or in hexane?
2. Look at the results of the reactions of chlorine water in test tubes 1 and 2. Did chlorine react with sodium bromide? With sodium iodide? What evidence do you have for any reaction that occurred?
3. Look at the results of the reactions of bromine water in test tubes 3 and 4. Did bromine react with sodium chloride? With sodium iodide? What evidence do you have for any reaction that occurred?
4. Look at the results of the reactions of iodine water in test tubes 5 and 6. Did iodine react with sodium chloride? With sodium bromide? What evidence do you have for any reaction that occurred?
5. Using Equation 1 in the *Background* section as a guide, write a balanced chemical equation for each reaction that occurred in test tubes 1–6.
6. Based on laboratory results, which halogen is most reactive? Least reactive? Explain.
7. Using lab data, write the activity series for the halogens, from most active to least active.
8. Assuming other nonmetals behave the same as the halogens, write a general statement that describes the periodic trend in the reactivity of nonmetals as you go down a column in the Periodic Table.
9. Use your answer to Question #8 to predict which Group VIA element, oxygen or sulfur, should be more reactive.
10. What effect in general did silver nitrate have on each of the sodium halide solutions?
11. What effect did ammonia have on each precipitate?
12. Write a general statement that describes the similarities and differences in the solubility behavior of silver halides.
13. *(Optional)* Write the chemical equation for the double replacement reaction of silver nitrate with each of the three sodium halide solutions. Label the solid product with (s) and write the name and color of each precipitate.

Teacher Notes

All in the Family

Microscale

Master Materials List *(for a class of 30 students working in pairs)*

Ammonia water, NH_3, 1.0 M, 90 mL
Bromine water, Br_2 in H_2O, 15 mL
Chlorine water, Cl_2 in H_2O, 15 mL
Iodine solution, I_2, 0.05 M, 15 mL
Hexane (organic solvent), C_6H_{14}, 90 mL*
Silver nitrate solution, $AgNO_3$, 0.1 M, 45 mL
Sodium bromide solution, NaBr, 0.1 M, 30 mL
Sodium chloride solution, NaCl, 0.1 M, 30 mL
Sodium iodide solution, NaI, 0.1 M, 30 mL
Marking pencil or pen, 15
Pipets, Beral-type
Test tubes, 12 × 75 mm, 90
Test tube rack, 15
Stirring rods, 15
Stoppers for test tubes, 90
Water, distilled or deionized

*The recommended solvent is "hexanes," a mixture of C_6H_{14} isomers.

Preparation of Solutions *(for a class of 30 students working in pairs)*

Halogen Water Solutions. It is best to prepare these solutions the day of lab. Avoid storing chlorine and bromine water—prepare only as much as is needed. Carry out the reactions in the hood. Label the solutions properly.

Chlorine Water: To prepare 30 mL of chlorine water, mix 5 mL of 5% sodium hypochlorite solution (bleach) with 10 mL of 1 M HCl. Dilute to 30 mL with distilled water.

Bromine Water: To prepare 30 mL of bromine water, mix 10 mL of 1 M NaBr and 10 mL of 1 M HCl solutions. Add 5 mL of 5% sodium hypochlorite solution (bleach), followed by 5 mL of distilled water.

The following solutions may be prepared ahead of time as convenient.

Iodine Solution, 0.05 M: Note: Elemental iodine is only slightly soluble in water. Potassium iodide is added to dissolve the iodine in the concentration needed. Dissolve 1 g of potassium iodide in about 30 mL of distilled water. Add 0.65 g of iodine and mix to dissolve. Dilute to 50 mL with distilled water.

Sodium chloride, 0.1 M. Add 0.3 g of sodium chloride to about 60 mL of water. Stir to dissolve and dilute to 50 mL with distilled water.

Sodium bromide, 1 M: (For bromine water preparation.) Add 5.2 g of sodium bromide to about 30 mL of water. Stir to dissolve, then dilute to 50 mL with distilled water.

Sodium Bromide, 0.1 M: (For Parts B and C.) Dilute 10 mL of freshly prepared 1 M solution to 100 mL with distilled water.

Sodium Iodide, 0.1 M: Add 0.75 g of sodium iodide to about 30 mL of water. Stir to dissolve and dilute to 50 mL with distilled water.

Silver Nitrate, 0.1 M: Add 0.85 g of silver nitrate to about 25 mL of water. Stir to dissolve and dilute to 50 mL with distilled water.

Safety Precautions

Conduct this experiment in a well-ventilated laboratory. If possible, dispense the halogen water solutions in the hood. Bromine water and chlorine water are toxic by ingestion and inhalation. Do not breathe bromine and chlorine vapors. Bromine and iodine water are skin and eye irritants. Avoid contact of all chemicals with eyes and skin. Hexane is a flammable organic liquid. Do not allow any flames in the laboratory during this activity. Wear chemical splash goggles and chemical-resistant gloves and apron. Consult current Material Safety Data Sheets for additional safety, handling, and disposal information.

Disposal

Consult your current *Flinn Scientific Catalog/Reference Manual* for general guidelines and specific procedures governing the disposal of laboratory waste. Chlorine water may be disposed of by allowing the chlorine vapor to escape from an open container under a hood in a well-ventilated, protected area. Bromine and iodine water solutions may be disposed of by reduction with sodium thiosulfate, according to Flinn Suggested Disposal Method #12a. Hexane may be disposed of according to Flinn Suggested Disposal Method #18a. Silver waste solutions may be disposed of by precipitating silver chloride according to Flinn Suggested Disposal Method #11. Silver chloride solid may be packaged for landfill disposal according to Flinn Suggested Disposal Method #26a.

Lab Hints

- Saturated chlorine and bromine water solutions are available from Flinn Scientific. Less concentrated solutions, however, are safe and easy to prepare and will also work. See the *Preparation of Solutions* section. If you elect to purchase ready-made solutions, purchase only the amounts needed for one school year. These solutions do not store well.
- Prior to lab, review with students the recommended safety precautions for working with halogens. Caution the students to never, ever sniff the halogen water solutions. Some students are tempted to do this with chlorine water, because it is colorless and the presence of chlorine is not obvious.
- The small amounts of chlorine and bromine water used in this experiment may be safely worked with at laboratory benches in a well-ventilated lab. If a hood is available, dispense the halogen water solutions in the hood. The students should stopper the halogen water tubes in the hood before carefully transferring them back to their lab benches. To avoid congestion when dispensing the reagents, stagger the starting points for Parts A/B and C. Half the student groups can start with Part A, the other half with Part C, then switch.
- To improve safety, set up two waste beakers in central locations, preferably in the hood, for immediate collection of laboratory waste: (1) a "Halogen Waste" beaker containing 100 mL of 50% sodium thiosulfate solution for Parts A and B; and (2) a "Silver Waste" beaker for Part C.
- The recommended organic solvent for Part A is "hexanes," a flammable but nontoxic mixture of n-hexane and other C_6H_{14} isomers. Many teachers prefer to use mineral oil instead of hexanes. Mineral oil is a safer substitute, but it is difficult to completely wash out of a test tube. Cyclohexane is another good solvent choice, although it is more expensive than hexane and slightly toxic. Although non-flammable, halogenated solvents are not recommended because they are toxic, possible carcinogens, and environmental hazards.

Teacher Notes

In Part B caution students not to shake the test tubes. Some vapor pressure may build up due to the hexane, and the stoppers may pop out of the test tubes.

Teacher Notes

Teaching Tips

- The placement of the periodic table in the curriculum varies from one textbook to another. The topic is usually introduced after atomic and electron structure have been covered, but before oxidation and reduction have been introduced. To avoid possible confusion, the terms oxidation and oxidizing agent have not been used in this experiment to describe the chemical reactivity of the halogens. Teachers who have covered the basics of oxidation and reduction may want to incorporate these concepts in their discussion of the chemical properties of the halogens. Fluorine is the strongest oxidizing agent among the halogens, while iodide ion is the strongest reducing agent among the halide anions.

- Depending on the sequence of topics in your curriculum, some students may not be prepared to write balanced chemical equations for all reactions that are observed in this lab. Using the material presented in the *Background* section, students should be able to write equations for the reactions of halogens with sodium halide salts. The double-replacement reactions of sodium halides and silver nitrate, and the subsequent reaction of silver chloride with ammonia, may not be familiar to students. Adjust the grading guidelines for the post-lab discussion questions based on the material that has been covered in your classroom. All students should be able to identify the periodic trends in the reactions of the halogens and the solubility of silver halides.

- The discovery of fluorine makes an interesting story at this time. As students discover by extension in this experiment, fluorine must be the most reactive of the halogens. In fact, it is the most reactive of all the nonmetals—the strongest oxidizing agent known—and extremely poisonous. For these reasons, fluorine was the last of the halogens to be discovered in its free element form. Chlorine, bromine, and iodine had all been isolated as their free elements by 1830. The element fluorine, however, was not successfully prepared until 1886 by the French chemist Henri Moissan. In the period between 1830 and 1886, several scientists died in their unsuccessful attempts to prepare elemental fluorine, and many more scientists, including Moissan, suffered serious health effects due to fluorine poisoning.

Answers to Pre-Lab Questions

1. Read the entire procedure and the recommended safety precautions. What hazards are associated with the use of halogens in the lab? How can these hazards be avoided?

   Chlorine and bromine water are toxic by inhalation and ingestion. Work with these solutions in a well-ventilated lab only. Never breathe chlorine and bromine. Bromine and iodine water are eye and skin irritants. Avoid contact of all chemicals with eyes and skin. Wear chemical splash goggles and chemical-resistant gloves and apron.

Students may notice that the silver halides appear to be light-sensitive, darkening on exposure to light. This observation provides a good starting point for talking about the applications of silver halides in photography.

2. Chlorine is used in water treatment plants and in swimming pools. Why is chlorine added to drinking water and to swimming pool water? Based on your answer to Question #1, suggest a possible disadvantage to the use of chlorine in water treatment.

   Chlorine is added as a disinfectant to water. Because of its high chemical reactivity, chlorine is used to oxidize and thereby kill bacteria and other microbial life forms. The chief disadvantage to the use of chlorine in water treatment is that chlorine is toxic—exposure to large concentrations of chlorine vapor is potentially lethal. ***Note to teacher:*** *There is also a growing awareness of the environmental hazard associated with the use of chlorine. Because it is so reactive, chlorine reacts with organic molecules, including those in our bodies, to form organochlorine compounds, which are also toxic and can accumulate in fish and wildlife.*

3. The following table shows the boiling points of the halogens. Describe in words the trend that is observed.

Halogen	F_2	Cl_2	Br_2	I_2
Boiling point	–188 °C	–34 °C	58 °C	184 °C

   The boiling points of the halogens increase in a regular and predictable manner as their molecular weights increase, due to the increasing size of the X_2 molecules. ***Note to teacher:*** *As the size of the molecules increases, there are more electrons and a greater surface area for attraction of the molecules by dispersion forces. The greater the attractive forces between the molecules, the harder it is to convert the molecules from the liquid phase to the gas phase, and the higher the boiling point of the liquid.*

Teacher Notes

Sample Data

Student data will vary.

Data Table A. *Color and Appearance of the Halogens*

Tubes	Halogen	Initial observations	Observations with hexane
1, 2	**Cl_2**	Colorless solution	Two liquid layers. Both layers are clear and colorless.
3, 4	**Br_2**	Light orange/yellow solution	Two liquid layers. Top layer is orange; bottom layer is yellow.
5, 6	**I_2**	Dark orange/brown solution	Two liquid layers. Top layer is purple; bottom layer is orange.

Data Table B. *Reactions of Halogens and Sodium Halides*

Tube	Halogen	Sodium Halide	Observations
1	**Cl_2**	**NaBr**	Top layer is orange; bottom layer is yellow.
2	**Cl_2**	**NaI**	Top layer is purple; bottom layer is orange.
3	**Br_2**	**NaCl**	NR (No change in colors of top and bottom layers.)
4	**Br_2**	**NaI**	Top layer is purple; bottom layer is light orange.
5	**I_2**	**NaCl**	NR (No change in colors of top and bottom layers.)
6	**I_2**	**NaBr**	NR (No change in colors of top and bottom layers.)

Data Table C. *Solubility of Metal Halides*

Tube	Sodium Halide	Observations with $AgNO_3$	Observations with aqueous ammonia
1	**NaCl**	White precipitate; clumpy and thick.	Precipitate breaks up and mostly dissolves. Solution is clear with trace amount of white-pink solid.
2	**NaBr**	Light cream precipitate; cloudy gray solution.	Cloudy gray suspension does not appear to dissolve much or settle fully. Light green solid at bottom.
3	**NaI**	Pale yellow precipitate; cloudy yellow solution.	Cloudy yellow suspension does not dissolve at all and does not settle.

Chlorine water may have a pale yellow tint. Hexane is less dense than water and is the upper layer in a two-phase water/hexane system.

Answers to Post-Lab Questions *(Student answers will vary.)*

1. How can each of the halogens—Cl_2, Br_2, and I_2—be detected in the aqueous layer and in the organic (hexane) layer? Based on your observations, are the halogens more soluble in water or in hexane?

   Chlorine is colorless in both water and in hexane solution. Bromine is yellow in water, darker orange in hexane solution. Iodine is dark orange/brown in water, purple in hexane. The halogens appear to be more soluble in hexane than in water based on the color changes observed when hexane is added to the halogen water solutions.

2. Look at the results of the reactions of chlorine water in test tubes 1 and 2. Did chlorine react with sodium bromide? With sodium iodide? What evidence do you have for any reaction that occurred?

   Chlorine reacted with both sodium bromide and sodium iodide. Evidence for reaction was the appearance of an orange color (elemental bromine) in the top organic layer when chlorine was mixed with sodium bromide and appearance of a purple color (elemental iodine) in the top organic layer when chlorine was mixed with sodium iodide.

3. Look at the results of the reactions of bromine water in test tubes 3 and 4. Did bromine react with sodium chloride? With sodium iodide? What evidence do you have for any reaction that occurred?

   Bromine reacted with sodium iodide but not with sodium chloride. Evidence for reaction with iodide ion was the appearance of a purple color (elemental iodine) in the top organic layer when bromine was mixed with sodium iodide. No change was observed in the colors of the top and bottom layers when bromine was mixed with sodium chloride.

4. Look at the results of the reactions of iodine water in test tubes 5 and 6. Did iodine react with sodium chloride? With sodium bromide? What evidence do you have for any reaction that occurred?

   Iodine did not react with either sodium chloride or sodium bromide. No changes were observed in the original colors of the top and bottom layers when the sodium halide salts were added.

5. Using Equation 1 in the *Background* section as a guide, write a balanced chemical equation for each double replacement reaction that occurred in test tubes 1–6.

   $2NaBr + Cl_2 \rightarrow Br_2 + 2NaCl$ *Test tube #1*

   $2NaI + Cl_2 \rightarrow I_2 + 2NaCl$ *Test tube #2*

   $2NaI + Br_2 \rightarrow I_2 + 2NaBr$ *Test tube #4*

Teacher Notes

6. Based on laboratory results, which halogen is most reactive? Least reactive? Explain.

   Chlorine is the most reactive of the halogens, because it reacted with both sodium bromide and sodium iodide. Iodine is the least reactive of the halogens, because it did not react with either sodium chloride or sodium bromide.

7. Using lab data, write the activity series for the halogens, from highest to lowest.

   Chlorine, bromine, iodine.

8. Write a general statement that describes the periodic trend in the reactivity of nonmetals as you go down a column in the Periodic Table.

   The reactivity of nonmetals in their free element forms decreases as you go down a column in the periodic table.

9. Use your answer to Question #8 to predict which Group VIA element, oxygen or sulfur, should be more reactive?

   The reactivity of nonmetals decreases as you go down a column in the periodic table. Based on this trend, oxygen should be more reactive than sulfur.

10. What effect does silver nitrate have on each of the sodium halide solutions?

    The chemical reaction between each of the sodium halide solutions and silver nitrate produces a solid precipitate, showing that silver salts of chloride, bromide, and iodide are insoluble in water.

11. What effect did ammonia have on each precipitate?

    Insoluble silver chloride dissolved when ammonia water was added, due to the formation of a silver-ammonia complex ion. Neither silver bromide nor silver iodide dissolved when ammonia water was added.

12. Write a general statement that describes the similarities and differences in the solubility behavior of silver halides.

    The silver halide salts—silver chloride, silver bromide, and silver iodide—are insoluble in water. Adding ammonia to suspensions of the silver halides dissolves silver chloride, but does not dissolve either silver bromide or silver iodide.

13. *(Optional)* Write the chemical equation for the reaction of silver nitrate with each of the three halide salt solutions. Label the solid product with (s) and write the name and color of each precipitate.

$NaCl(aq) + AgNO_3(aq) \rightarrow AgCl(s) + NaNO_3(aq)$
Silver chloride
(white)

$NaBr(aq) + AgNO_3(aq) \rightarrow AgBr(s) + NaNO_3(aq)$
Silver bromide
(light tan)

$NaI(aq) + AgNO_3(aq) \rightarrow AgI(s) + NaNO_3(aq)$
Silver iodide
(yellow)

Teacher Notes

Periodic Activity of Metals
Chemical Demonstration

Introduction

The elements are classified based on similarities, differences, and trends in their properties, including their chemical reactions. The reactions of alkali and alkaline earth metals with water are some of the most dramatic and spectacular of all chemical reactions. Mixtures bubble and boil, fizz and hiss, and may even smoke and burn. Introduce the study of the periodic table and periodic trends with this exciting demonstration of the activity of metals.

Concepts

- Periodic table
- Alkali and alkaline earth metals
- Physical and chemical properties
- Periodic trends
- Metal activity
- Acid–base indicator

Materials

Calcium turnings, Ca, 0.2–0.5 g
Lithium metal, Li, 0.1–0.5 g
Magnesium ribbon, Mg, 3-cm strip
Sodium metal, Na, 0.1–0.5 g
Phenolphthalein, 0.5% solution, 2 mL
Water, distilled or deionized, 600 mL
Beaker, Berzelius (tall-form), Pyrex®, 400-mL, 4
Petri dishes, Pyrex®, 4
ChemCam™ Video Camera (optional)

Safety Precautions

Lithium and sodium are flammable, water-reactive, corrosive solids; dangerous when exposed to heat or flame. They react violenty with water to produce flammable hydrogen gas and solutions of corrosive metal hydroxides. Hydrogen gas may be released in sufficient quantities to cause ignition or explosion. These metals are shipped in dry mineral oil. Store them in mineral oil until immediately before use. Do not allow these metals to stand exposed to air from one class period to another or for extended periods of time. Purchasing small, pre-cut pieces of sodium greatly reduces its potential hazard. Calcium metal is flammable in finely divided form and reacts upon contact with water to give flammable hydrogen gas and corrosive calcium hydroxide. Magnesium metal is a flammable solid and burns with an intense flame. Do not handle any of the metals with bare hands. Wear chemical splash goggles, chemical-resistant gloves, and a chemical-resistant apron. All students or spectators should also be wearing chemical splash goggles during this demonstration. Use a Class D fire extinguisher like Flinn Met-L-X® or have a generous supply of dry sand to use as a fire extinguisher. Consult current Material Safety Data Sheets for safety, handling, and disposal information.

Preparation

1. Obtain four 400-mL beakers and label them Li, Na, Mg, and Ca. Add approximately 150 mL of distilled or deionized water to each beaker.
2. Label four Petri dishes Li, Na, Mg, and Ca and place them on the demonstration table.

Teacher Notes

3. Obtain or prepare small metal pieces suitable for demonstrations. Metal pieces should be approximately 0.5 cm × 0.5 cm × 0.5 cm and their mass should not exceed 0.5 g. Use a 3-cm strip of magnesium ribbon.

Procedure

1. Place each metal in its respective Petri dish and ask the class to observe and compare the physical properties of the metals: color, luster (shine), and hardness.
2. Ask student volunteers to list the properties of the elements on the chalkboard for later reference.
3. *(Optional)* Rank the metals with respect to their physical properties. Which metal appears to be the shiniest? Softest (or hardest)?
4. Use forceps or tongs to quickly transfer one piece of sodium metal to the appropriately labeled beaker. Have students make detailed observations: the metal, which initially floats on the surface, immediately begins to "hiss and sizzle" on the water surface and appears to melt (from the heat that is released). Popping sounds are heard and the metal begins to bounce around and finally disappear. A smoky gas forms and briefly ignites on the surface of the water.
5. Repeat step 4 using lithium metal instead of sodium. As observations are made, ask students to compare the rate and intensity of the reaction with that of sodium.
6. Repeat step 4 twice more, using calcium and magnesium, respectively. Compare the activity of each metal against the previous metal and against sodium as a reference metal.
7. Refer to the detailed observations to rank the metals in order of activity. Which metal is most active? Least active?
8. Write the symbols of the metals on the board in the same arrangement of rows and columns as they are found on the periodic table.
9. Based on class discussion, draw arrows across the rows and up or down the columns to indicate the direction in which metal activity increases in the periodic table, as shown below.

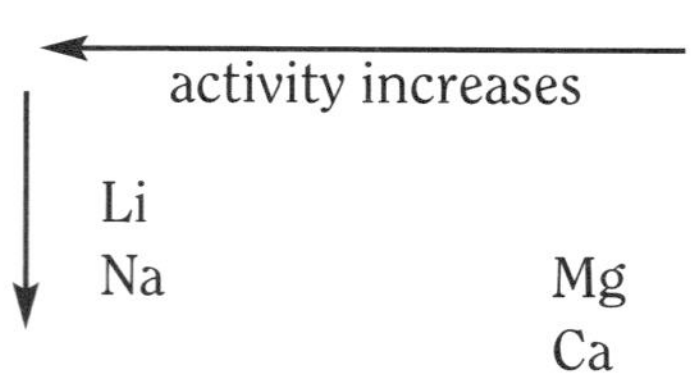

10. Ask students to predict the activity of potassium metal based on the observed periodic trend in the activity of metals. Discuss the extreme reactivity of potassium metal (and why it was not used in this demonstration).
11. After the metals have reacted (with the exception of magnesium, which does not react under these conditions), add 5 drops of phenolphthalein solution to each beaker.

Teacher Notes

12. Note the color change to red-violet. What does the color change "indicate?" What types of compounds give this color change? Discuss the possible identity of the product(s).

13. Write balanced chemical equations for the reactions of the metals with water. Discuss the evidence for the formation of both hydrogen gas and metal hydroxides.

14. *(Optional)* In conclusion, ask students to write a paragraph describing in words the physical and chemical properties of one of the metals. *Example:*

    "Sodium is a soft, silver-white solid. Upon exposure to air it gradually develops a white oxide coating. It can be cut with a knife. It is less dense than water and reacts spontaneously and vigorously on contact with water. The metal piece appears to pop or sizzle on the surface and a smoky white gas forms. The metal may ignite on the surface of the water in the vicinity of the smoke. The products are hydrogen gas and sodium hydroxide. The hydrogen gas that is formed 'pops' and briefly ignites. Sodium hydroxide makes the solution basic (red) to phenolphthalein indicator."

Disposal

Use tongs or forceps to remove unreacted magnesium from its beaker. Dispose of excess magnesium metal according to Flinn Suggested Disposal Method #26a. Do not dispose of any of the other reaction mixtures until all of the metal in each has completely reacted. The resulting basic solutions can be neutralized and disposed of according to Flinn Suggested Disposal Method #10.

Tips

- We strongly recommend the purchase of small quantities only of lithium and sodium metal. Purchase only the amounts that will be used in one academic year. Sodium is available in small, precut pieces that are suitable for demonstrations.
- We do not recommend the use of potassium in the high school science laboratory. Potassium is considerably more water-reactive than sodium and thus poses a greater fire and explosion hazard. In addition, potassium poses a significant *and often undetectable* explosion hazard because of its propensity to form a superoxide (peroxide) coating on its surface. Potassium reacts with oxygen in air to form a coating of yellow potassium superoxide (KO_2), even when the metal is stored under dry mineral oil. Old pieces of potassium are thus extremely dangerous. When the metal is cut, the pressure of the knife may touch off a violent, uncontrollable, explosive reaction between the superoxide coating and the underlying metal.
- Calcium metal must be reasonably fresh to react with water. Old (dull) calcium metal does not react. Check calcium turnings for a dull white coating that easily rubs off.
- The use of a ChemCam™ video camera makes it easier for students to observe the appearance and properties of the metal pieces.
- Demonstrate the softness of lithium and sodium metals by showing they can be cut with a spatula or knife.

Teacher Notes

- In ranking the metals in order of their activity, it is easier to begin with pairwise comparisons. Which is more active—sodium or lithium? Calcium or magnesium? Magnesium or sodium?
- Does activity of metals correlate with their hardness? With density? The answer, a firm "maybe," depends on the comparisons being made. The alkali metals as a group are softer and less dense than their alkaline earth metal nearest neighbors, and also more reactive. Within the group of alkali metals, however, the opposite trend is observed. Lithium is less dense but also less reactive than sodium.

Discussion

Sodium reacts with water to form hydrogen gas and sodium hydroxide, according to the following balanced chemical equation.

$$2Na(s) + 2H_2O(l) \rightarrow H_2(g) + 2NaOH(aq)$$

As sodium metal reacts with water, a great deal of heat is generated. The sodium melts and seems to float and bob on the water surface. The oxide coating which may have dulled the metal disappears and its silvery gray, metallic character is more apparent. The evolution of hydrogen gas is evident in the production of a white smoke, which may pop and ignite as it is heated above its flash point. The formation of sodium hydroxide, a strong base, is inferred from the color change observed with phenolphthalein, an acid–base indicator. Phenolphthalein is colorless in neutral or slightly basic solutions ($pH < 8$) and red in more basic solutions ($pH > 10$). Between pH 8 and 10 phenolphthalein appears various shades of pink.

The balanced chemical equations for reactions of other active metals with water are given below.

$$2Li(s) + 2H_2O(l) \rightarrow H_2(g) + 2LiOH(aq)$$

$$Ca(s) + 2H_2O(l) \rightarrow H_2(g) + Ca(OH)_2(aq)$$

Of the four metals tested, sodium is the most active and magnesium is the least active. Magnesium does not react with water under these conditions (it may react slightly in hot water). The order of metal reactivity is Na > Li, Ca >>Mg. Periodic trends in the activity of metals are generally attributed to differences in their ionization energies. The activity of metals increases as the value of their first ionization energy decreases. Within a vertical column (group or family) of elements in the periodic table, ionization energy decreases from top to bottom. As a result, metal activity *increases* going down a column in the periodic table (K > Na > Li; Ca > Mg). Across a horizontal row (period or series) in the periodic table, ionization energy increases from left to right. As a result, metal activity *decreases* from left to right across a row in the periodic table (Na >> Mg; K >> Ca).

Teacher Notes

Safe Swimming with Sodium
Chemical Demonstration

Introduction

No chemistry class is complete without the spectacular demonstration of alkali metals reacting with water. This demonstration is a novel variation that is much safer to perform than the traditional sodium demonstration of simply dropping a small piece of sodium into a beaker of water.

Concepts

- Alkali metals
- Metal hydroxides
- Acid–base indicator
- Density

Materials

Sodium metal, Na, 1 small piece
Mineral oil, 200 mL
Phenolphthalein, 0.5% solution, 5 drops
Water, 200 mL
Hydrometer cylinder, 600-mL, or graduated cylinder
Ring stand and clamp
Lithium (optional), Li, 1 small piece

Safety Precautions

Sodium metal is a flammable, corrosive solid; dangerous when exposed to heat or flame; dangerous by reaction with moist air, water, or any oxidizer. Purchasing small, pre-cut pieces of sodium greatly reduces its potential hazard. Sodium reacts with water to produce flammable hydrogen gas and a solution of corrosive sodium hydroxide. Wear chemical splash goggles, chemical-resistant gloves, and a chemical-resistant apron. Consult current Material Safety Data Sheets for additional safety, handling, and disposal information.

Consider purchasing sodium in pre-cut, ready-to-use form. Flinn Catalog No. S0329 includes five small pieces of sodium (0.3–0.5 g each) that are perfect for demonstrations. The sodium is packaged under mineral oil in a PVC-coated glass bottle. The bottle is placed in a plastic bag and then a metal can for safer storage. See your current Flinn Scientific Catalog/Reference Manual for a photo of the sodium packaging and the Safe Swimming with Sodium demonstration.

Procedure

1. Clamp a hydrometer cylinder or large graduated cylinder to a ring stand for support.
2. Add about 200 mL of water to the cylinder along with a few drops of phenolphthalein solution.
3. Add 200 mL of mineral oil, forming a layer above the water.
4. Drop a piece of sodium, about the size of a kernel of corn, into the cylinder and observe the reaction.

Disposal

Consult your current *Flinn Scientific Catalog/Reference Manual* for both general guidelines and specific procedures governing the disposal of laboratory waste. Do not dispose of anything until the sodium has completely reacted. The hydrocarbon layer can be stored and reused for future demonstrations and labs. The aqueous solution can be flushed down the drain with excess water according to Flinn Suggested Disposal Method #26a.

Teacher Notes

Tips

- Various hydrocarbon solvents work well. Light mineral oil has low volatility and odor but high viscosity, so the sodium "swims" for a longer time. Lighter hydrocarbon mixtures, such as kerosene, turpentine, xylenes, or mineral spirits, have a lower viscosity but a higher volatility and odor and present a greater fire risk if the apparatus is knocked over.
- If the mineral oil is wet, the sodium metal may react more quickly at first. The heat generated by this reaction will cause the sodium to melt and become porous. This decreases the density of sodium metal so that it floats in the mineral oil layer.
- The colorless water-phenolphthalein layer can be regenerated by the addition of a small amount of dilute acid, such as 1 M HCl, after the reaction with sodium is complete. The demonstration apparatus can thus be used several times during the day.
- Hydrogen gas produced in this reaction can be collected using standard gas collection procedures and safety precautions.

Discussion

When sodium metal is added to the cylinder, it sinks through the mineral oil layer until it reaches the interface between the water and mineral oil layers. When sodium metal comes in contact with the water layer, it spontaneously reacts to form hydrogen gas and sodium hydroxide, according to the following balanced chemical equation:

$$2Na(s) + 2H_2O(l) \rightarrow H_2(g) + 2NaOH(aq)$$

The evolution of hydrogen gas is immediately evident and hydrogen bubbles adhering to the sodium "lift" or carry it back up into the hydrocarbon layer. This temporarily stops the reaction. When the bubbles adhering to the metal surface disperse, the piece of sodium dives back down to the water-hydrocarbon interface, where it reacts again and then "swims" back up into the hydrocarbon layer. This behavior continues until the reaction is complete. The amount of hydrogen and heat evolved is kept under control by the "swimming" behavior, making this demonstration quite safe. During the reaction, the piece of sodium is largely devoid of corrosion, allowing the students to view its silvery gray, metallic appearance. The aqueous layer contains phenolphthalein and turns pink due to the production of basic sodium hydroxide. No precipitate is formed, which shows that sodium hydroxide is soluble in water.

Density is an important physical property that can be used to separate materials or control reactions. Sodium has a density of 0.97 g/mL and sits at the interface of water and a hydrocarbon solvent. In contrast, lithium has a density of 0.54 g/mL and will float on top of the hydrocarbon layer. (Try it!) The interface between two immiscible solvents is an effective site for controlling chemical reactions.

Teacher Notes

Plotting Trends
A Periodic Table Activity

Introduction

Most high school chemistry textbooks report a wealth of numerical data to identify periodic trends in the properties of elements. Ionization energies, atomic radii, electronegativity values, and electron affinities—all are dutifully tabulated and graphed. But what do all the numbers mean? Students cannot measure these quantities, they cannot see their relative size, they cannot feel their relative effects. In this cooperative activity, students use microscale reaction plates and straws of different lengths to construct three-dimensional bar-type charts of the physical properties of the elements. The resulting charts are visually impressive and clearly reveal to students the meaning of periodic trends in the properties of elements.

Concepts

- Periodic table
- Periodic trends
- Electron affinity
- Ionization energy
- Atomic radius
- Electronegativity

Materials *(for class of 28 students working in groups of 4)*

Calculator, at least 1 per student group

Handout of representative element data, 28

Index cards, 4 × 6 inches, 7

Microscale reaction plates, 96-well (8 × 12 layout), 7

Periodic table (for reference), at least 1 per student group

Straws, 350 (50 per student group)

Scissors, at least 1 per student group

Metric rulers, marked in millimeters, at least 1 per student group

Safety Precautions

Although the materials in this activity are considered nonhazardous, follow all normal laboratory safety guidelines.

Procedure

1. Form a working group with three other students.
2. Obtain a handout of representative element data, a microscale reaction plate, a metric ruler, scissors, and 50 plastic straws.
3. Find your assigned physical property on the handout of element data for 42 representative (main group) elements.

Teacher Notes

4. Find the maximum value of the assigned physical property for the elements listed. *Example:* The maximum value of the density for the elements listed is 11.342 g/cm^3 (for lead).

5. Let the length of the straw minus one cm represent this maximum value. This will be the scale for all of the other measurements of that property. *Example:* For a straw that is 19.5 cm long, a straw length of 18.5 cm would represent a density of 11.342 g/cm^3. The scale is thus 18.5 cm = 11.342 g/cm^3. Round off straw lengths to 0.1 cm (1 mm).

6. Using this "straw" scale as a ratio, calculate the straw length that is needed to represent the assigned property for each element in the list. *Example:* The density of beryllium is 1.848 g/cm^3. Solving Equation 1 for the straw length (sl) shows that a straw length of 3.0 cm would represent the density of beryllium. Round off all straw lengths to 0.1 cm.

$$\frac{18.5\text{ (cm)}}{11.342} = \frac{sl\text{ (cm)}}{1.848} \qquad sl = (18.5 \times 1.848)/11.342 = 3.0\text{ cm} \qquad \textit{Equation 1}$$

7. Add 1 cm to the calculated straw length for each element and cut a straw to that length. *Example:* Cut a straw 4.0 cm long to represent beryllium.

8. Place the straw in the reaction plate according to its position in the periodic table. Remember, the transition elements and lanthanides and actinides are not included in the list of representative elements. *Example:* Beryllium is placed in row 2, column 2.

9. Repeat steps 6–8 for each element in the list.

10. Determine the nature of any periodic relationship or trend that may exist for the assigned property of the elements.

11. Propose an explanation for the observed trend.

12. Create a descriptive card to be displayed with the three-dimensional chart. Include the following information on the descriptive card:
    - Names of group members
    - The assigned physical property of the element
    - Description of the observed trend
    - Proposed explanation for the observed trend

The wells in the well plate are 1 cm deep. The cut straw lengths are therefore determined by adding 1 cm to the straw length calculated in step 5. Students may need help at first visualizing the placement of straws in the reaction plate. Notice that row 1 (period 1) has only two elements, H and He, in columns 1 and 8, respectively.

Tips

- Make enough copies of the handout of representative element data so that each student obtains an individual copy. The handout lists the properties of 42 representative (main group) elements, corresponding to the Group IA–Group VIII elements in periods 1–6. The atomic numbers for these elements are 1–20, 31–38, 49–56, and 81–86. The transition metals and lanthanides and actinides are not included on this handout.

Teacher Notes

- Seven physical properties of the elements are included on the handout. This is ideal for a class of 28 students working in groups of four. Each group selects or is assigned a unique property of the elements—all physical properties on the handout are accounted for and there are no duplications. For smaller or larger class sizes, adjust the size of the working groups and discard or double-up on the assigned physical properties, as appropriate.
- This is a cooperative class activity. Monitor or assess student groups to be sure that students are sharing responsibilities and discussing questions or choices to arrive at consensus.
- Set aside a central, common area for students to display their charts and the accompanying descriptive cards. Students can then report on their investigation and findings to the class and lead a discussion on the proposed explanation for their observed trend.
- To save time, reduce the number of elements in the list to four rows (periods) of elements. For example, using elements with atomic numbers 1–20 and 31–36 reduces the number of elements from 42 to 26. All of the periodic trends will still reveal themselves with this reduced number of elements.
- A sample "straw chart" of atomic radius is illustrated below.

Not every property of the elements exhibits clear periodic trends. Ionization energy, electron affinity, electronegativity, and atomic radius show periodic trends both within a group and across a period. Density varies in a periodic pattern within a group and across a series if the gaseous elements are excluded. There is also a pattern in the melting points of the elements, but it is not as predictable as the density pattern. Atomic mass is not a periodic propety of the elements.

Atomic Radius Straw Chart

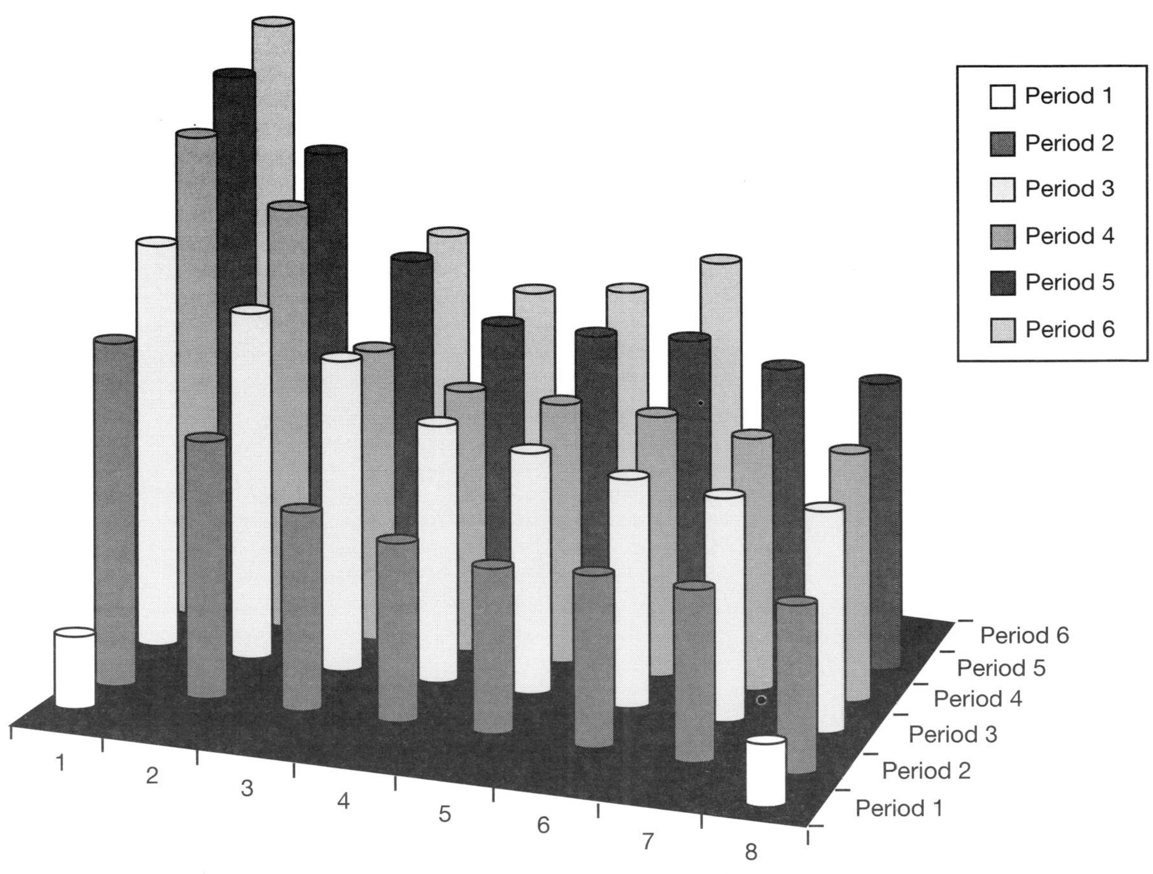

Teacher Notes

Representative Element Data

Symbol	Atomic Number	Atomic Mass (amu)	Density (g/cm^3)	Ionization Energy (kJ/mol)	Electron Affinity (kJ/mol)	Electro-Negativity	Atomic Radius (pm)	Melting Point (K)
H	1	1.008	0.0001	1312	72.8	2.2	32	13.4
He	2	4.003	0.0002	2372	–21		31	0.95
Li	3	6.941	0.534	520	59.6	0.98	152	454
Be	4	9.012	1.848	899	–241	1.57	112	1560
B	5	10.81	2.35	801	26.7	2.04	85	2349
C	6	12.011	2.266	1086	121.9	2.55	77	4188
N	7	14.007	0.0012	1402	0	3.04	70	63
O	8	15.999	0.0014	1314	141	3.44	73	55
F	9	18.998	0.0017	1681	328	3.98	72	54
Ne	10	20.18	0.0009	2081	–29		71	24
Na	11	22.99	0.968	496	52.9	0.93	186	371
Mg	12	24.305	1.738	738	–230	1.31	160	923
Al	13	26.981	2.699	578	42.5	1.61	143	934
Si	14	28.086	2.336	786	133.6	1.9	118	1685
P	15	30.974	1.823	1012	72	2.19	110	317
S	16	32.07	2.069	1000	200.4	2.58	103	380
Cl	17	35.453	0.0032	1251	349	3.16	100	172
Ar	18	39.948	0.0018	1521	–34		98	84
K	19	39.1	0.856	419	48.4	0.82	227	336
Ca	20	40.08	1.55	590	–156	1	197	1115
* 3d elements omitted								
Ga	31	69.72	5.904	579	28.9	1.81	135	303
Ge	32	72.61	5.323	762	119	2.01	122	1211
As	33	74.92	5.778	947	78.2	2.18	120	1090
Se	34	78.96	4.285	941	195	2.55	119	450
Br	35	79.9	3.1	1140	324.7	2.96	114	266
Kr	36	83.8	0.0037	1351	–39	3	112	116
Rb	37	85.47	1.532	403	46.9	0.82	248	312
Sr	38	87.62	2.63	550	–167	0.95	215	1030
* 4d elements omitted								
In	49	114.82	7.31	558	28.9	1.78	167	430
Sn	50	118.71	7.265	709	107.3	1.88	141	505
Sb	51	121.75	6.697	834	103.2	2.05	140	904
Te	52	127.6	6.25	869	190.2	2.1	142	723
I	53	126.91	4.94	1008	295.2	2.66	133	387
Xe	54	131.29	0.0059	1170	–40	2.6	131	161
Cs	55	132.91	1.9	376	45.5	0.79	265	302
Ba	56	137.33	3.62	503	–52	0.89	222	1000
* 5d elements omitted								
* 4f elements omitted								
Tl	81	204.38	11.85	589	19.3	1.83	170	577
Pb	82	207.2	11.342	716	35.1	2.1	146	601
Bi	83	208.98	9.808	703	91.3	2.02	150	545
Po	84	210	9.142	812	183.3	2	168	527
At	85	210		890	270.2	2.2		302
Rn	86	222	0.0097	1037	–41			202

*Transition metals and inner transition elements are not included in this table.

Representative element data were taken from the "Periodic Table" video disk published by the Journal of Chemical Education and from the Flinn Scientific Periodic Table.

Teacher Notes

The Ultimate Element Crossword Puzzle

At the bottom of this page is an alphabetical listing of the first 103 elements. Sixty-two of them have been made into a crossword puzzle below. Use the clues provided to fill in the appropriate letters. The clues themselves contain over 70 word plays. Circle all the ones you find. For example, in one of the clues the words "our scenic" are used and refer to the answer to that clue: "arsenic."

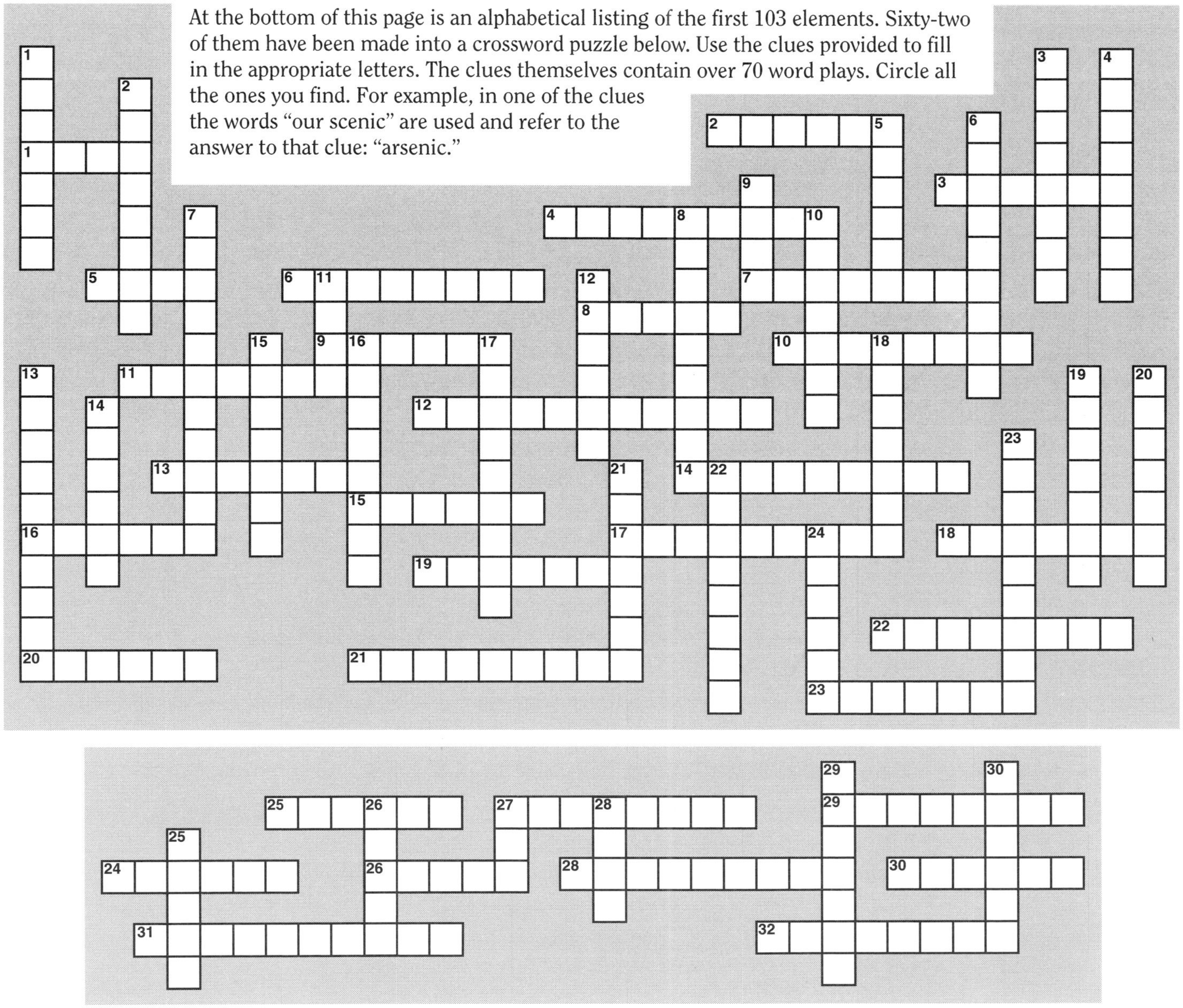

actinium
aluminum
americium
antimony
argon
arsenic
astatine
barium
berkelium
beryllium
bismuth
boron
bromine
cadmium
calcium
californium
carbon
cerium
cesium
chlorine
chromium
cobalt
copper
curium
dysprosium
einsteinium
erbium
europium
fermium
fluorine
francium
gadolinium
gallium
germanium
gold
hafnium
helium
holmium
hydrogen
indium
iodine
iridium
iron
krypton
lanthanum
lawrencium
lead
lithium
lutetium
magnesium
manganese
mendelevium
mercury
molybdenum
neodymium
neon
neptunium
nickel
niobium
nitrogen
nobelium
osmium
oxygen
palladium
phosphorus
platinum
plutonium
polonium
potassium
praseodymium
promethium
protactinium
radium
radon
rhenium
rhodium
rubidium
ruthenium
samarium
scandium
selenium
silicon
silver
sodium
strontium
sulfur
tantalum
technetium
tellurium
terbium
thallium
thorium
thulium
tin
titanium
tungsten
uranium
vanadium
xenon
ytterbium
yttrium
zinc
zirconium

For extra credit, have students identify the word play or pun in each clue. "Cows see them"—calcium. "All you minimum"—aluminum.

Teacher Notes

ACROSS

1. In batteries, stained glass windows, and old paint; but not in pencils, as some might lead you to believe. Also known for its density, malleability (and toxicity).
2. Matches, acid rain, and a chigger repellent (and originally associated with eternal damnation), so far as we know, its molecules are square-dance-shaped.
3. Pipes, wires and Ms. Liberty's skin, whether you're a seasoned cop or a brand new rookie, a penny's worth isn't what it used to be.
4. In bananas, salt subs, & of course, special K cereal, but this alkali metal is also used in lavender fireworks.
5. Anyone can see that, despite its total disregard for others, this member of the noble gas family has a bright future in advertising—I'd put money on it!
6. Fertilizers, explosives & cryogenics, we come in contact with this gas more than any other. Long ago, at night, Trojan men would breathe in a mixture containing 80% of this gas, before going into battle!
7. Crystals of its salts are known for their crisp colors, this metal puts the shine on your car bumper.
8. Used in strobe lamps, this "stranger" may be a member of the royal family, but recently he has been seen on dates with such commoners as F and O!
9. Used as an alloy in ball-point pens, nearly twice as dense as lead (if Dorothy's house were made of this metal, she never would have gotten anywhere!), the strong odor given off by this metal is a result of its highly toxic oxide!
10. Not for sale at Arbee's, this alkali is a real go-getter in vacuum tubes and is known for its ruby-colored flame test.
11. Greenish yellow in color, used in bleach, drinking water, PVC pipes (that's logical) or in WWI as the world's first war gas.
12. This relatively recent addition to the table does not occur naturally, it was discovered in the all burnt-up debris analyzed from the first H-bomb explosion (the energy of which was equal to the mass lost times the speed of light squared).
13. Make your silly list of pros and cons: Bill Gates, Cinderella & some top-notch super models all owe their good fortune to this metalloid.
14. Used in flares, flash bulbs and magnificent incendiary bombs, but also in mag wheels & in M.O.M.
15. This shiny nonmetal gives off purple vapors; at blood donor centers, this goes on before the needle goes in, and in Ohio, diner and restaurant owners add it to their salt.
16. Without this noble lightweight, the Goodyear Company would never have gotten off the ground, and scuba divers would really be hurting (and doctors might be unable to heal them).
17. One of the most reactive rare earth metals, (seriously) and often used in lighter flints and carbon-arc lamps, but using this toxic element in Heinz steak sauce would not be a good plan (the number of protons has little to do with an element's uses!)
18. Its orange-red spectral line is how the modern day meter is defined, and—not to sound cryptic—but its fictitious ore is the Achilles' heel for the man of steel.
19. Highly toxic, as are its compounds, including its oxide which has the distinct odor of garlic. Also used in transistors: think about that as you drive along our scenic highways listening to your radio!
20. When the Lone Ranger photographs his horse's fillings in the mirror he must wonder why this #1 metal (in terms of electrical and heat conductivity), is still considered second best by most athletes.
21. An alloying agent in steel & an active ingredient in poison ivy lotion, its silicate may be cubic but it won't last forever (sorry, girls . . .).
22. Originally thrown away as a "fool's silver," now considered more valuable than gold, especially by recording artists (and DJ's who spin their "platters").
23. Lighten up, today, this low density metal is used in batteries, tomorrow, it may be used in the matter–antimatter chamber of the U.S.S. Enterprise!
24. Although this alkaline earth metal is very toxic, I know a doctor who often asks his patients to drink a nice thick shake of its sulfate (and he rarely ends up having to bury them).
25. Has your car been acting up? You may want to check the antifreeze or oil or gasoline or vinyl seat covers or . . . yourself!
26. This great-great-great-granddaughter of U-238 is a dense, noble and silent killer, but do not be afraid. On the other hand, you may want to sleep with your windows open.
27. WOW, This wondrous worldly stone must weigh a ton (hence the Swedish name), but this weak wire filament is so light!
28. This most deadly poison was used in the Apollo program to power equipment on the lunar surface. Critical mass of this fissionable isotope can blast you to the outer limits of the solar system!
29. This group V metalloid is quite versatile: from tracer bullets to fireproofing to infrared detection. Not that I'm against cash or anything, but I should just pay off my ex-wife with this valuable stuff each month.
30. One of the most metallic metals, named for its sky-blue spectral lines, atomic clocks using this element are accurate to 5 sec in 300 years . . . or 1 sec in 60 years (that's easy math).
31. Just ten protons short of a full deck, welders know this steel-strengthening transition metal as "Molly B. Denim."
32. Albeit the most abundant metal in the Earth's crust, yet so difficult to extract from its ore that this shiny lightweight used to be more valuable than gold. Now everyone (even all you minimum waste advocates) treat it like trash!

Teacher Notes

DOWN

1. It must have taken some gall in 1871 for Mendeleev to predict the existence of this low-melting metal. Then, four years later "eka-aluminum" was discovered by a Gallic chemist.
2. You're number one in our books and in the universe for that matter, the ultimate fuel and building block, but don't try to hide your genuinely explosive nature: two of the worst aviation accidents in U.S. history resulted from your reaction with oxygen.
3. One might think that Fr, Ge and Sc might somehow be alloyed together to make this metal, but no, it's an element by itself, with an atomic mass of 151.96.
4. This pale yellow gas is considered the most reactive of all the elements. Even water will ignite in its presence. And after 70 years of continuous work, chemists were finally able to isolate this element (I bet that made them smile).
5. Used for years in self-luminous paint for glow-in-the-dark clock faces and in radiological treatment for cancer. One atom of this alkaline earth metal has as many protons as there are keys on a piano!
6. Its name is derived from the Greek neos, meaning "new" and didymos, meaning "twin," because for years no one could separate it from its next-door neighbor on the periodic table.
7. Named after the long-bearded Russian himself (who was not an MD, but without whom there would be no periodic table). Like all transuranium elements, #101 is synthesized (not found in nature) and it has no stable isotopes.
8. Similar in nature to the other alkaline earth metals (which together make up a pretty strong team in terms of their reactivity). Salts of this metal are used in flares and fireworks producing a beautiful crimson color.
9. Has more applications than you might think: from solders to makeup to printer's ink; paints, plastics, and fencing (chain-link), and yes, of course, the kitchen ___.
10. The only metal that is a liquid at ordinary temperatures (ordinary for Earth, that is, not a hotter planet!), nicknamed "quick silver," used in thermometers, barometers & electrical switches, very dense, also very toxic (hard to imagine a more curious metal).
11. Some might consider it ironic that the same metal that's used in car bodies & Ferris wheels is also used to fortify our breakfast cereal. The most abundant & important metal on Earth, it's in our blood to treasure it.
12. To metals and non-metals alike, this element bonds like epoxy—generally in a very exothermic manner (in other words, it makes things burn), but, oh, what trouble we would be in without these atoms, (both the pairs down here and the trios up there).
13. Discovered in 1669 by Brandt who prepared it from urine (but that's not how it got its symbol!). This non-metal is used in matches, fertilizers and detergents.
14. This transition metal is especially known for the deep blue color its salts produce when added to glass and ceramics (as produced by the Goblin Porcelain Co., Baltimore, MD, for example).
15. Ubiquitous & unusual, named after the 7th rock from the sun, one lb of this nuclear fuel source is equivalent to 1500 tons of coal, you have to respect that!
16. The black form of this nonmetal conducts electricity better when light is shined upon it, so it is used in Xerox toner, thus without this element you can cancel any immediate plans to photocopy this page!
17. Man can easily survive (and so can woman) without Ag or Au, but take away this trace dietary mineral and you are in eminent risk of a vitamin B deficiency!
18. Irredeemably the most corrosion resistant metal, #77 was alloyed with Pt to make the standard meter bar of Paris.
19. Why is this metal used in magnets, low melting alloys, & pink antidiarrheal medication? I make it my business to know!
20. This halogen is a liquid at room temperature, and though its symbol may remind you of the cold, its compounds serve as flameproofing agents!
21. Dairy farmers see milking machines as quick extractors of this alkaline earth metal, but I bet the cows see them differently!
22. Nothing is as effective as stating the obvious, but with less than one ounce of this halogen existing on the Earth, this element isn't really where it's at!
23. Being a lightweight with a high melting point, this group II metal is an ideal hardening agent in alloys. Originally, chemists tested for this element by its sweet taste, but its high toxicity makes this not such a brilliant idea!
24. Used in nichrome wire, alnico magnets, stainless steel barnacle scrapers(!), and as a catalyst for hydrogenating vegetable oil—the kind you might buy at the local five and dime.
25. A rather inactive gas, this royal family member makes up nearly 1% of our atmosphere, it is used as the inert gas in lightbulbs (when N & O are gone, the filament lasts a lot longer!)
26. Hamlet's soliloquy pondered whether or not this metalloid was diatomic! The nitride of this element can be as hard as diamond, and can be used for drill bits (talk about boring!).
27. Snips can cut this metal into tiny pieces, used for solders (and soldiers), and coated onto steel cans (for beans, tennis balls, etc.) to prevent corrosion.
28. "HEY YOU! Come back here with my first place medal!" This king of metals is so malleable that a single ounce can be beaten out into a 300 ft^2 sheet 1,000 times thinner than paper, and so inert that jewelry dating back to ancient times still looks as good as new.
29. This close relative of Zn is used with nickel to make Nicad rechargeable batteries. It is famous for its yellow paint pigment, but it is very toxic and is one of the many poisons found in cigarette smoke.
30. Naturally, this most abundant alkali metal is used in soaps, glass and pretzels. Most carbonated drinks, however, report having very low levels of this element—which seems so dumb when you think about it.

Answers to the Ultimate Element Crossword Puzzle

Teacher Notes

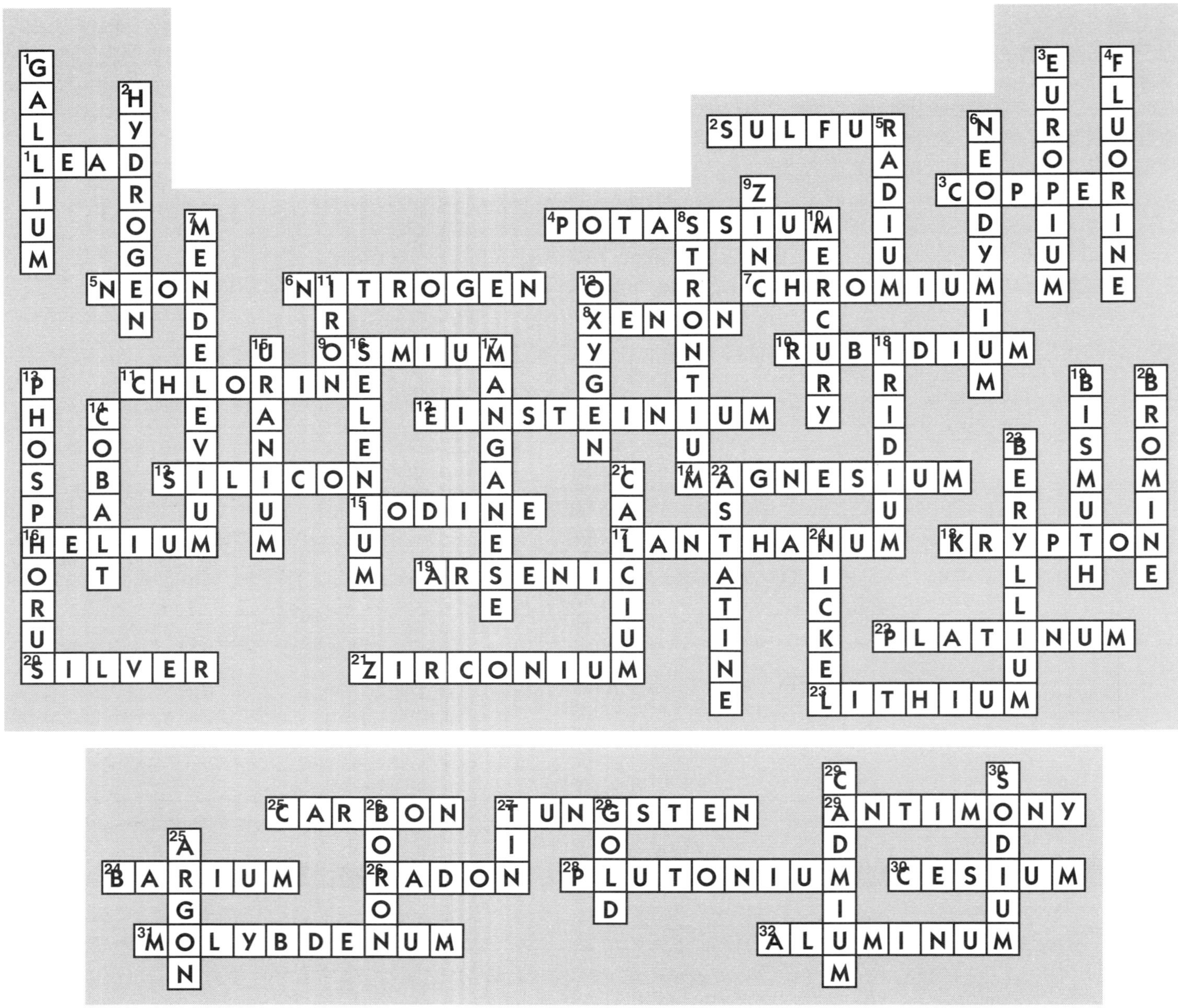

Teacher Notes

Solubility Patterns
Chemical Demonstration

Microscale

Introduction

The alkaline earth metals are so-named because their oxides are highly basic (alkaline) and because they occur abundantly on Earth. Alkaline earth metal compounds—carbonates, sulfates, silicates—are abundant in minerals such as *dolomite* (calcium and magnesium carbonate), *epsomite* (magnesium sulfate), and *baryte* (barium sulfate). The solubilities of these compounds vary widely, depending on the metal cation. Are there any patterns or periodic trends in the solubility behavior of alkaline earth metal compounds?

Concepts

- Alkaline earth metals
- Double-replacement reactions
- Periodic trends
- Ionic compounds

Materials

Alkaline earth metal chlorides

Barium chloride, 0.1 M, $BaCl_2$, 6 mL
Calcium chloride, 0.1 M, $CaCl_2$, 6 mL
Magnesium chloride, 0.1 M, $MgCl_2$, 6 mL
Strontium chloride, 0.1 M, $SrCl_2$, 6 mL
Pipets, Beral-type, graduated, 9
Reaction plate, 24-well, 1

Testing Solutions

Ammonium hydroxide, 6 M, NH_4OH, 4 mL
Ammonium oxalate, 0.25 M, $(NH_4)_2C_2O_4$, 4 mL
Potassium chromate, 0.5 M, K_2CrO_4, 4 mL
Sodium carbonate, 1 M, Na_2CO_3, 4 mL
Sodium sulfate, 1 M, $(Na)_2SO_4$, 4 mL

Safety Precautions

Ammonium hydroxide is toxic by ingestion and inhalation. Both the liquid and vapor are highly irritating to the eyes and the respiratory tract. Avoid breathing ammonium hydroxide vapor. Conduct this demonstratin in a well-ventilated laboratory. Potassium chromate is highly toxic and is corrosive to skin, eyes, and the respiratory tract. Strontium and barium compounds are toxic by ingestion. Oxalates are toxic by ingestion and are irritating to body tissues. Avoid contact of all chemicals with eyes and skin. Wear chemical splash goggles and chemical-resistant gloves and apron. Please consult current Material Safety Data Sheets for additional safety, handling, and disposal information.

Procedure

1. Place a 24-well reaction plate onto the stage of the overhead projector and turn on the projector. Note that each well is identified by a unique combination of a letter and a number, where the letters A–D refer to horizontal rows and the numbers 1–6 refer to vertical columns.

Teacher Notes

2. Using a clean, graduated Beral-type pipet for each solution, add 1 mL (about 25 drops) of alkaline earth metal chloride solution to each well in a horizontal row, as follows:
   - magnesium chloride to wells A1–A6
   - calcium chloride to wells B1–B6
   - strontium chloride to wells C1–C6
   - barium chloride to wells D1–D6
3. Using a clean, graduated Beral-type pipet for each solution, add 1 mL (about 25 drops) of testing solution to each well in a vertical column, as follows:
   - ammonium hydroxide to wells A1–D1
   - potassium chromate to wells A2–D2
   - sodium sulfate to wells A3–D3
   - sodium carbonate to wells A4–D4
   - ammonium oxalate to wells A5–D5

   Note: The last column serves as a control to identify the absence of a precipitate.
4. Have students draw a table of circles on the chalkboard in the same layout as the wells in the reaction plate. Identify each horizontal row with the correct alkaline earth metal chloride and each vertical column with the correct testing solution.
5. Observe the reactions that develop in the reaction plate and have students record the results in the table of circles on the chalkboard. Use the abbreviations PPT and NR to note the formation of a precipitate or no reaction, respectively.
6. What patterns or trends are obvious in the solubility behavior of the alkaline earth metal compounds? Which alkaline earth metal ion formed the most precipitates? The fewest? Which testing solution gave the most precipitates? The fewest?
7. Ask students to identify any periodic trend in the solubility behavior of alkaline earth metal compounds. Is there any relationship between the solubility of alkaline earth metal compounds and the position of the metal in the periodic table? Are there any exceptions to the general solubility pattern?
8. *Fill in the blank in the following sentence to describe the general solubility trend:* The solubility of alkaline earth metal compounds decreases as you proceed ______________ in the periodic table.
9. Propose an explanation for the observed solubility pattern.
10. Use the observed solubility pattern to predict a chemical method for the separation of a mixture of calcium and barium ions in solution. (Imagine a solution that is 0.1 M in both $CaCl_2$ and $BaCl_2$. What reagents can be added to this mixture and in what order to separate the two compounds?)

Disposal

Consult your current *Flinn Scientific Catalog/Reference Manual* for general guidelines and specific procedures governing the disposal of laboratory waste. Barium compounds may be disposed of according to Flinn Suggested Disposal Method #27h. Chromates may be disposed of according to Flinn Suggested Disposal Method #12a. All other solutions can be flushed down the drain with excess water according to Flinn Suggested Disposal Method #26b.

Alkaline earth metal nitrates may be used instead of the chlorides. No procedural changes are needed if this substitution is made.

Teacher Notes

Tips

- This demonstration provides an excellent exercise for writing and balancing chemical equations and net ionic equations.
- The demonstration can be extended to include the identification of an unknown. Carry an unknown solution containing one or two different alkaline earth metal cations through the sequence of precipitation reactions. Use the resulting solubility pattern to identify the unknown cation(s).
- Discuss the relationship between the solubility pattern of alkaline earth metal compounds and hard water. Hard water contains relatively high concentrations of magnesium and calcium ions. The problems caused by hard water range from a nuisance (soaps leave soap scum, calcium stearate; detergents are not effective) to costly industrial "boiler scale" water treatment programs (to eliminate the formation of calcium carbonate).

Discussion

Periodic trends are observed in the solubility of alkaline earth metal compounds. Although their chlorides and nitrates are all water-soluble, alkaline earth metal compounds with other anions do not always dissolve in water. The solubility of alkaline earth metal compounds with different anions is tested by carrying out *double replacement reactions*. Reaction of calcium chloride with sodium carbonate, for example, leads to an exchange of anions between the two metals to give calcium carbonate, which is insoluble in water and precipitates out as a solid when the two solutions are mixed. The chemical equation for this reaction is given in Equation 1, where the abbreviations (aq) and (s) refer to aqueous solutions and solid precipitates, respectively.

$CaCl_2(aq)$	+	$Na_2CO_3(aq)$	$\rightarrow$	$CaCO_3(s)$	+	$2NaCl(aq)$	*Equation 1*
calcium chloride		sodium carbonate		calcium carbonate		sodium chloride	

The solubility pattern observed for alkaline earth metal compounds is shown below. The solubility of alkaline earth metal compounds decreases as you go down the column in the periodic table, i.e., solubility decreases as the atomic mass of the alkaline earth metal increases.

	OH^-	CrO_4^{2-}	SO_4^{2-}	CO_3^{2-}	$C_2O_4^{2-}$	control
$MgCl_2$	A1 PPT	A2 NR	A3 NR	A4 NR	A5 NR	A6
$CaCl_2$	B1 NR	B2 NR	B3 NR	B4 PPT	B5 PPT	B6
$SrCl_2$	C1 NR	C2 NR	C3 PPT	C4 PPT	C5 PPT	C6
$BaCl_2$	D1 PPT	D2 PPT	D3 PPT	D4 PPT	D5 PPT	D6

In general, magnesium formed the fewest precipitates, while barium formed the most precipitates. Magnesium hydroxide is an exception to the general solubility trend of alkaline earth metal compounds.

Safety and Disposal Guidelines

Safety Guidelines

Teachers owe their students a duty of care to protect them from harm and to take reasonable precautions to prevent accidents from occurring. A teacher's duty of care includes the following:

- Supervising students in the classroom.
- Providing adequate instructions for students to perform the tasks required of them.
- Warning students of the possible dangers involved in performing the activity.
- Providing safe facilities and equipment for the performance of the activity.
- Maintaining laboratory equipment in proper working order.

Safety Contract

The first step in creating a safe laboratory environment is to develop a safety contract that describes the rules of the laboratory for your students. Before a student ever sets foot in a laboratory, the safety contract should be reviewed and then signed by the student and a parent or guardian. Please contact Flinn Scientific at 800-452-1261 or visit the Flinn Website at www.flinnsci.com to request a free copy of the Flinn Scientific Safety Contract.

To fulfill your duty of care, observe the following guidelines:

1. **Be prepared.** Practice all experiments and demonstrations beforehand. Never perform a lab activity if you have not tested it, if you do not understand it, or if you do not have the resources to perform it safely.

2. **Set a good example.** The teacher is the most visible and important role model. Wear your safety goggles whenever you are working in the lab, even (or especially) when class is not in session. Students learn from your good example—whether you are preparing reagents, testing a procedure, or performing a demonstration.

3. **Maintain a safe lab environment.** Provide high-quality goggles that offer adequate protection and are comfortable to wear. Make sure there is proper safety equipment in the laboratory and that it is maintained in good working order. Inspect all safety equipment on a regular basis to ensure its readiness.

4. **Start with safety.** Incorporate safety into each laboratory exercise. Begin each lab period with a discussion of the properties of the chemicals or procedures used in the experiment and any special precautions—including goggle use—that must be observed. Pre-lab assignments are an ideal mechanism to ensure to ensure that students are prepared for lab and understand the safety precautions. Record all safety instruction in your lesson plan.

5. **Proper instruction.** Demonstrate new or unusual laboratory procedures before every activity. Instruct students on the safe way to handle chemicals, glassware, and equipment.

6. **Supervision.** Never leave students unattended—always provide adequate supervision. Work with school administrators to make sure that class size does not exceed the capacity of the room or your ability to maintain a safe lab environment. Be prepared and alert to what students are doing so that you can prevent accidents before they happen.

7. **Understand your resources.** Know yourself, your students, and your resources. Use discretion in choosing experiments and demonstrations that match your background and fit within the knowledge and skill level of your students and the resources of your classroom. You are the best judge of what will work or not. Do not perform any activities that you feel are unsafe, that you are uncomfortable performing, or that you do not have the proper equipment for.

Safety Precautions

Specific safety precautions have been written for every experiment and demonstration in this book. The safety information describes the hazardous nature of each chemical and the specific precautions that must be followed to avoid exposure or accidents. The safety section also alerts you to potential dangers in the procedure or techniques. Regardless of what lab program you use, it is important to maintain a library of current Material Safety Data Sheets for all chemicals in your inventory. Please consult current MSDS for additional safety, handling, and disposal information.

Disposal Procedures

The disposal procedures included in this book are based on the Suggested Laboratory Chemical Disposal Procedures found in the *Flinn Scientific Catalog/Reference Manual*. The disposal procedures are only suggestions—do not use these procedures without first consulting with your local government regulatory officials.

Many of the experiments and demonstrations produce small volumes of aqueous solutions that can be flushed down the drain with excess water. Do not use this procedure if your drains empty into groundwater through a septic system or into a storm sewer. Local regulations may be more strict on drain disposal than the practices suggested in this book and in the *Flinn Scientific Catalog/Reference Manual*. You must determine what types of disposal procedures are permitted in your area—contact your local authorities.

Any suggested disposal method that includes "discard in the trash" requires your active attention and involvement. Make sure that the material is no longer reactive, is placed in a suitable container (plastic bag or bottle), and is in accordance with local landfill regulations. Please do not inadvertently perform any extra "demonstrations" due to unpredictable chemical reactions occurring in your trash can. Think before you throw!

Finally, please read all the narratives before you attempt any Suggested Laboratory Chemical Disposal Procedure found in your current *Flinn Scientific Catalog/Reference Manual*.

Flinn Scientific is your most trusted and reliable source of reference, safety and disposal information for all chemicals used in the high school science lab. To request a complementary copy of the most recent *Flinn Scientific Catalog/Reference Manual,* call us at 800-452-1261 or visit our website at www.flinnsci.com.

Flinn Scientific's Student Safety Contract

PURPOSE

Science is a hands-on laboratory class. You will be doing many laboratory activities which require the use of hazardous chemicals. Safety in the science classroom is the #1 priority for students, teachers, and parents. To ensure a safe science classroom, a list of rules has been developed and provided to you in this student safety contract. These rules must be followed at all times. Two copies of the contract are provided. One copy must be signed by both you and a parent or guardian before you can participate in the laboratory. The second copy is to be kept in your science notebook as a constant reminder of the safety rules.

GENERAL RULES

1. Conduct yourself in a responsible manner at all times in the laboratory.
2. Follow all written and verbal instructions carefully. If you do not understand a direction or part of a procedure, ask the instructor before proceeding.
3. Never work alone. No student may work in the laboratory without an instructor present.
4. When first entering a science room, do not touch any equipment, chemicals, or other materials in the laboratory area until you are instructed to do so.
5. Do not eat food, drink beverages, or chew gum in the laboratory. Do not use laboratory glassware as containers for food or beverages.
6. Perform only those experiments authorized by the instructor. Never do anything in the laboratory that is not called for in the laboratory procedures or by your instructor. Carefully follow all instructions, both written and oral. Unauthorized experiments are prohibited.
7. Be prepared for your work in the laboratory. Read all procedures thoroughly before entering the laboratory.
8. Never fool around in the laboratory. Horseplay, practical jokes, and pranks are dangerous and prohibited.
9. Observe good housekeeping practices. Work areas should be kept clean and tidy at all times. Bring only your laboratory instructions, worksheets, and/or reports to the work area. Other materials (books, purses, backpacks, etc.) should be stored in the classroom area.
10. Keep aisles clear. Push your chair under the desk when not in use.
11. Know the locations and operating procedures of all safety equipment including the first aid kit, eyewash station, safety shower, fire extinguisher, and fire blanket. Know where the fire alarm and the exits are located.
12. Always work in a well-ventilated area. Use the fume hood when working with volatile substances or poisonous vapors. Never place your head into the fume hood.
13. Be alert and proceed with caution at all times in the laboratory. Notify the instructor immediately of any unsafe conditions you observe.
14. Dispose of all chemical waste properly. Never mix chemicals in sink drains. Sinks are to be used only for water and those solutions designated by the instructor. Solid chemicals, metals, matches, filter paper, and all other insoluble materials are to be disposed of in the proper waste containers, not in the sink. Check the label of all waste containers twice before adding your chemical waste to the container.
15. Labels and equipment instructions must be read carefully before use. Set up and use the prescribed apparatus as directed in the laboratory instructions or by your instructor.
16. Keep hands away from face, eyes, mouth and body while using chemicals or preserved specimens. Wash your hands with soap and water after performing all experiments. Clean all work surfaces and apparatus at the end of the experiment. Return all equipment clean and in working order to the proper storage area.
17. Experiments must be personally monitored at all times. You will be assigned a laboratory station at which to work. Do not wander around the room, distract other students, or interfere with the laboratory experiments of others.
18. Students are never permitted in the science storage rooms or preparation areas unless given specific permission by their instructor.
19. Know what to do if there is a fire drill during a laboratory period; containers must be closed, gas valves turned off, fume hoods turned off, and any electrical equipment turned off.
20. Handle all living organisms used in a laboratory activity in a humane manner. Preserved biological materials are to be treated with respect and disposed of properly.
21. When using knives and other sharp instruments, always carry with tips and points pointing down and away. Always cut away from your body. Never try to catch falling sharp instruments. Grasp sharp instruments only by the handles.
22. If you have a medical condition (e.g., allergies, pregnancy, etc.), check with your physician prior to working in lab.

CLOTHING

23. Any time chemicals, heat, or glassware are used, students will wear laboratory goggles. There will be no exceptions to this rule!
24. Contact lenses should not be worn in the laboratory unless you have permission from your instructor.
25. Dress properly during a laboratory activity. Long hair, dangling jewelry, and loose or baggy clothing are a hazard in the laboratory. Long hair must be tied back and dangling jewelry and loose or baggy clothing must be secured. Shoes must completely cover the foot. No sandals allowed.
26. Lab aprons have been provided for your use and should be worn during laboratory activities.

ACCIDENTS AND INJURIES

27. Report any accident (spill, breakage, etc.) or injury (cut, burn, etc.) to the instructor immediately, no matter how trivial it may appear.
28. If you or your lab partner are hurt, immediately yell out "Code one, Code one" to get the instructor's attention.
29. If a chemical splashes in your eye(s) or on your skin, immediately flush with running water from the eyewash station or safety shower for at least 20 minutes. Notify the instructor immediately.
30. When mercury thermometers are broken, mercury must not be touched. Notify the instructor immediately.

HANDLING CHEMICALS

31. All chemicals in the laboratory are to be considered dangerous. Do not touch, taste, or smell any chemicals unless specifically instructed to do so. The proper technique for smelling chemical fumes will be demonstrated to you.
32. Check the label on chemical bottles twice before removing any of the contents. Take only as much chemical as you need.
33. Never return unused chemicals to their original containers.

Flinn Scientific's Student Safety Contract

34. Never use mouth suction to fill a pipet. Use a rubber bulb or pipet pump.
35. When transferring reagents from one container to another, hold the containers away from your body.
36. Acids must be handled with extreme care. You will be shown the proper method for diluting strong acids. Always add acid to water, swirl or stir the solution and be careful of the heat produced, particularly with sulfuric acid.
37. Handle flammable hazardous liquids over a pan to contain spills. Never dispense flammable liquids anywhere near an open flame or source of heat.
38. Never remove chemicals or other materials from the laboratory area.
39. Take great care when transporting acids and other chemicals from one part of the laboratory to another. Hold them securely and walk carefully.

HANDLING GLASSWARE AND EQUIPMENT

40. Carry glass tubing, especially long pieces, in a vertical position to minimize the likelihood of breakage and injury.
41. Never handle broken glass with your bare hands. Use a brush and dustpan to clean up broken glass. Place broken or waste glassware in the designated glass disposal container.
42. Inserting and removing glass tubing from rubber stoppers can be dangerous. Always lubricate glassware (tubing, thistle tubes, thermometers, etc.) before attempting to insert it in a stopper. Always protect your hands with towels or cotton gloves when inserting glass tubing into, or removing it from, a rubber stopper. If a piece of glassware becomes "frozen" in a stopper, take it to your instructor for removal.
43. Fill wash bottles only with distilled water and use only as intended, e.g., rinsing glassware and equipment, or adding water to a container.
44. When removing an electrical plug from its socket, grasp the plug, not the electrical cord. Hands must be completely dry before touching an electrical switch, plug, or outlet.
45. Examine glassware before each use. Never use chipped or cracked glassware. Never use dirty glassware.
46. Report damaged electrical equipment immediately. Look for things such as frayed cords, exposed wires, and loose connections. Do not use damaged electrical equipment.
47. If you do not understand how to use a piece of equipment, ask the instructor for help.
48. Do not immerse hot glassware in cold water; it may shatter.

HEATING SUBSTANCES

49. Exercise extreme caution when using a gas burner. Take care that hair, clothing and hands are a safe distance from the flame at all times. Do not put any substance into the flame unless specifically instructed to do so. Never reach over an exposed flame. Light gas (or alcohol) burners only as instructed by the teacher.
50. Never leave a lit burner unattended. Never leave anything that is being heated or is visibly reacting unattended. Always turn the burner or hot plate off when not in use.
51. You will be instructed in the proper method of heating and boiling liquids in test tubes. Do not point the open end of a test tube being heated at yourself or anyone else.
52. Heated metals and glass remain very hot for a long time. They should be set aside to cool and picked up with caution. Use tongs or heat-protective gloves if necessary.
53. Never look into a container that is being heated.
54. Do not place hot apparatus directly on the laboratory desk. Always use an insulating pad. Allow plenty of time for hot apparatus to cool before touching it.
55. When bending glass, allow time for the glass to cool before further handling. Hot and cold glass have the same visual appearance. Determine if an object is hot by bringing the back of your hand close to it prior to grasping it.

QUESTIONS

56. Do you wear contact lenses?
    ☐ YES ☐ NO
57. Are you color blind?
    ☐ YES ☐ NO
58. Do you have allergies?
    ☐ YES ☐ NO

    If so, list specific allergies ____________

    ____

    ____

AGREEMENT

I, ______________________________ , (student's name) have read and agree to follow all of the safety rules set forth in this contract. I realize that I must obey these rules to ensure my own safety, and that of my fellow students and instructors. I will cooperate to the fullest extent with my instructor and fellow students to maintain a safe lab environment. I will also closely follow the oral and written instructions provided by the instructor. I am aware that any violation of this safety contract that results in unsafe conduct in the laboratory or misbehavior on my part, may result in being removed from the laboratory, detention, receiving a failing grade, and/or dismissal from the course.

____
Student Signature

____
Date

Dear Parent or Guardian:

We feel that you should be informed regarding the school's effort to create and maintain a safe science classroom/laboratory environment.

With the cooperation of the instructors, parents, and students, a safety instruction program can eliminate, prevent, and correct possible hazards.

You should be aware of the safety instructions your son/daughter will receive before engaging in any laboratory work. Please read the list of safety rules above. No student will be permitted to perform laboratory activities unless this contract is signed by both the student and parent/guardian and is on file with the teacher.

Your signature on this contract indicates that you have read this Student Safety Contract, are aware of the measures taken to ensure the safety of your son/daughter in the science laboratory, and will instruct your son/daughter to uphold his/her agreement to follow these rules and procedures in the laboratory.

____
Parent/Guardian Signature

____
Date

Content Standards

Experiments and Demonstrations

Content Standards	It's in the Cards	Density Is a Periodic Property	Periodic Trends and Properites of Elements	All in the Family	Periodic Activity of Metals	Safe Swimming with Sodium	Plotting Trends	Ultimate Element Crossword Puzzle	Solubility Patterns
Unifying Concepts and Processes									
Systems, order, and organization	✓	✓	✓				✓		
Evidence, models, and explanation	✓	✓	✓	✓	✓		✓		✓
Constancy, change, and measurement		✓					✓		
Evolution and equilibrium									
Form and function									
Science as Inquiry									
Identify questions and concepts that guide scientific investigation	✓	✓	✓	✓	✓				
Design and conduct scientific investigations		✓	✓	✓	✓				✓
Use technology and mathematics to improve scientific investigations		✓					✓		
Formulate and revise scientific explanations and models using logic and evidence	✓		✓	✓	✓		✓		✓
Recognize and analyze alternative explanations and models	✓						✓		
Communicate and defend a scientific argument							✓		
Understanding scientific inquiry	✓								
Physical Science									
Structure of atoms									
Structure and properties of matter	✓	✓	✓	✓	✓	✓	✓	✓	✓
Chemical reactions			✓	✓	✓	✓		✓	
Motions and forces									
Conservation of energy and the increase in disorder									
Interactions of energy and matter									

Content Standards *(continued)*

	Experiments and Demonstrations								
	It's in the Cards	Density Is a Periodic Property	Periodic Trends and Properites of Elements	All in the Family	Periodic Activity of Metals	Safe Swimming with Sodium	Plotting Trends	Ultimate Element Crossword Puzzle	Solubility Patterns
Science and Technology									
Identify a problem or design an opportunity							✓		
Propose designs and choose between alternative solutions									
Implement a proposed solution									
Evaluate the solution and its consequences									
Communicate the problem, process, and solution							✓		
Understand science and technology									
Science in Personal and Social Perspectives									
Personal and community health									
Population growth									
Natural resources									
Environmental quality									
Natural and human-induced hazards									
Science and technology in local, national, and global challenges									
History and Nature of Science									
Science as a human endeavor	✓	✓						✓	
Nature of scientific knowledge	✓	✓					✓		
Historical perspectives	✓	✓						✓	

(for a class of 30 students working in pairs)

Experiments and Demonstrations*

Chemicals	Flinn Scientific Catalog No.	It's in the Cards	Density Is a Periodic Property	Periodic Trends and Properites of Elements	All in the Family	Periodic Activity of Metals	Safe Swimming with Sodium	Plotting Trends	Ultimate Element Crossword Puzzle	Solubility Patterns
Aluminum foil	A0019			30 cm × 30 cm						
Ammonia water, 1.0 M	A0038				90 mL					
Ammonium hydroxide solution, 6 M	A0192									4 mL
Ammonium oxalate solution, 0.25 M	A0198									4 mL
Barium chloride solution, 0.1 M	B0143			45 mL						6 mL
—or—										
Barium chloride dihydrate	B0170			2.4 g						
Bromine water	B0166				15 mL					
Calcium chloride solution, 0.1 M	C0234			45 mL						6 mL
—or—										
Calcium chloride dihydrate	C0196			1.5 g						
Calcium turnings	C0345			8 g		0.5 g				
Chlorine water	C0055				15 mL					
Hexane	H0046				90 mL					
Hydrochloric acid, 0.5 M	H0035			45 mL						
Iodine solution, 0.05 M	I0148				15 mL					
Lead shot	L0008		500–600 g							
Lithium	L0057					0.5 g	optional			
Magnesium chloride solution, 0.1 M	M0121			45 mL						6 mL
—or—										
Magnesium chloride hexahydrate	M0146			2.0 g						
Magnesium ribbon	M0139			30 cm		3 cm				
Mineral oil	M0064						200 mL			
Phenolphthalein solution, 0.5%	P0115					2 mL	1 mL			
Potassium chromate solution, 0.5 M	P0162									4 mL
Potassium iodate solution, 0.2 M	P0168			75 mL						
—or—										
Potassium iodate powder	P0063			4.3 g						
Silicon lumps	S0013		120–150 g							
Silver nitrate solution, 0.1 M	S0305				45 mL					
—or—										
Silver nitrate	S0274				0.85 g					
Sodium bromide solution, 0.1 M	S0269				30 mL					
—or—										
Sodium bromide	S0215				5.2 g					
Sodium lumps	S0329					0.5 g	0.5 g			

*Many of these experiments and demonstrations will be adapted into student laboratory kits. Consult your current *Flinn Scientific Catalog/Reference Manual* for kit availability.

(for a class of 30 students working in pairs)

Experiments and Demonstrations

	Flinn Scientific Catalog No.	It's in the Cards	Density is a Periodic Property	Periodic Trends and Properites of Elements	All in the Family	Periodic Activity of Metals	Safe Swimming with Sodium	Plotting Trends	Ultimate Element Crossword Puzzle	Solubility Patterns
Chemicals, continued										
Sodium carbonate solution, 1 M —or— Sodium carbonate monohydrate	S0234 S0054			75 mL 12.4 g						4 mL
Sodium chloride solution, 0.1 M —or— Sodium chloride	S0237 S0061				30 mL 0.3 g					
Sodium iodide solution, 0.1 M —or— Sodium iodide	S0245 S0083				30 mL 0.75 g					
Sodium sulfate solution, 1 M —or— Sodium sulfate decahydrate	S0352 S0225			75 mL 32 g						4 mL
Strontium chloride solution, 0.1 M —or— Strontium chloride hexahydrate	S0255 S0357			45 mL 2.7 g						6 mL
Tin shot	T0016		400–500 g							
Glassware										
Beakers, 50-mL	GP1005		45							
Beakers, Pyrex®, Berzelius, 400-mL	GP1059					4				
Graduated cylinders, 25-mL	GP2010		15							
Hydrometer cylinder, 600-mL	AP8599						1			
Test tubes, 12 × 75 mm	GP6062				90					
Stirring rods	GP5075				15					
General Equipment and Miscellaneous										
Balance, centigram (0.01-g precision)	OB2059		3							
ChemCam™ Video Camera	AP4560					optional				
Forceps	AP8328		15	15						
Hot plate	AP8386		optional							
Laboratory Techniques Guide, pkg of 30	AP6248		1	1	1					
Litmus paper, red	AP7945			1 vial						
Matches	AP1935			optional						
Periodic Table of the Elements, notebook size, pkg of 50	AP9020		1	1	1			1	1	
Petri dishes, Pyrex®	AP3019					4				
Pipets, Beral-type	AP1721			150	135					9

Continued on next page

(for a class of 30 students working in pairs)

Experiments and Demonstrations

General Equipment and Miscellaneous, cont'd.	Flinn Scientific Catalog No.	It's in the Cards	Density is a Periodic Property	Periodic Trends and Properites of Elements	All in the Family	Periodic Activity of Metals	Safe Swimming with Sodium	Plotting Trends	Ultimate Element Crossword Puzzle	Solubility Patterns
Reaction plate, 24-well	AP1447			15						1
Reaction plate, 96-well	AP1448							7		
Ring, support with rod clamp, 3-inch	AP1320						1			
Ruler, metric	AP5386							7		
Scissors	AP5394							7		
Stoppers, cork, size 2	AP8302				90					
Straws, plastic, pkg of 500	AP6025							1		
Support stand	AP8228						1			
Test tube rack	AP1319				15					
Thermometer, digital	AP6049			15						
Tongs	AP8236		15							
Water, distilled or deionized	W0007			✓	✓	✓	✓			✓
Wax pencil*	AP8292		15		15					

*Or other pens for labeling

The Flinn Periodic Table

We asked the country's best high school chemistry teachers "What do you want in a periodic table?" We took their suggestions and incorporated them into Flinn's new two-sided, four-color periodic table.

Teachers asked for a periodic table where the most vital information—atomic number, atomic symbol, name, and atomic mass—doesn't get lost. We accomplished this by grouping this information on one side of each element box. We also included information like melting and boiling points, heat of fusion and vaporization, electronegativity, electron configurations, and much more.

Teachers also asked us for historical information for each element, common uses to generate student interest, derivation of the name or symbol of each element, and abundance of the element in crustal rocks.

The table is available three ways. First, as a 58″ × 42″ two-sided laminated wall chart with an aluminum top and bottom band for easy hanging. A second option is to purchase the front side only of our 58″ × 42″ periodic table mounted on a spring roller assembly. Printed on tough ChemDura stock, this chart will last for years. The third option is to purchase the 17″ × 11″ foldover notebook size. Perfect for student use. Packages of 50.

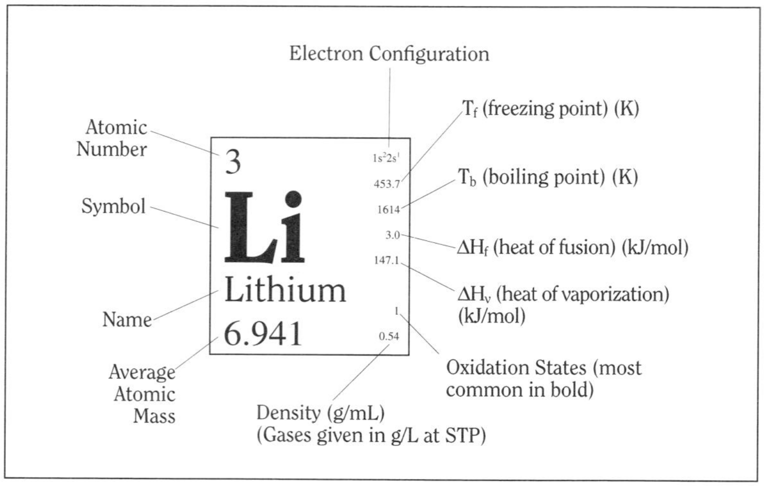

Catalog No.	Description	Price/Each
AP9021	Flinn Periodic Table, Two-Sided Hanging Wall Chart	Consult Your Current Flinn Catalog/ Reference Manual.
AP4530	Flinn Periodic Table, One-Sided Roller-Mounted Chart	
AP9020	Flinn Periodic Table, Notebook Size, Package of 50	

Periodic Table of the Elements

	IA	IIA	IIIB	IVB	VB	VIB	VIIB	VIIIB			IB	IIB	IIIA	IVA	VA	VIA	VIIA	VIIIA
	1	2	3	4	5	6	7	8	9	10	11	12	13	14	15	16	17	18
1	1 **H** Hydrogen 1.008																	2 **He** Helium 4.003
2	3 **Li** Lithium 6.94	4 **Be** Beryllium 9.01											5 **B** Boron 10.81	6 **C** Carbon 12.01	7 **N** Nitrogen 14.01	8 **O** Oxygen 16.00	9 **F** Fluorine 19.00	10 **Ne** Neon 20.18
3	11 **Na** Sodium 22.99	12 **Mg** Magnesium 24.31											13 **Al** Aluminum 26.98	14 **Si** Silicon 28.09	15 **P** Phosphorus 30.97	16 **S** Sulfur 32.07	17 **Cl** Chlorine 35.45	18 **Ar** Argon 39.95
4	19 **K** Potassium 39.10	20 **Ca** Calcium 40.08	21 **Sc** Scandium 44.96	22 **Ti** Titanium 47.88	23 **V** Vanadium 50.94	24 **Cr** Chromium 52.00	25 **Mn** Manganese 54.94	26 **Fe** Iron 55.85	27 **Co** Cobalt 58.93	28 **Ni** Nickel 58.69	29 **Cu** Copper 63.55	30 **Zn** Zinc 65.39	31 **Ga** Gallium 69.72	32 **Ge** Germanium 72.61	33 **As** Arsenic 74.92	34 **Se** Selenium 78.96	35 **Br** Bromine 79.90	36 **Kr** Krypton 83.80
5	37 **Rb** Rubidium 85.47	38 **Sr** Strontium 87.62	39 **Y** Yttrium 88.91	40 **Zr** Zirconium 91.22	41 **Nb** Niobium 92.91	42 **Mo** Molybdenum 95.94	43 **Tc** Technetium 98.91	44 **Ru** Ruthenium 101.07	45 **Rh** Rhodium 102.91	46 **Pd** Palladium 106.42	47 **Ag** Silver 107.87	48 **Cd** Cadmium 112.41	49 **In** Indium 114.82	50 **Sn** Tin 118.71	51 **Sb** Antimony 121.75	52 **Te** Tellurium 127.6	53 **I** Iodine 126.90	54 **Xe** Xenon 131.29
6	55 **Cs** Cesium 132.91	56 **Ba** Barium 137.33	57-71 **La-Lu** ★	72 **Hf** Hafnium 178.49	73 **Ta** Tantalum 180.95	74 **W** Tungsten 183.85	75 **Re** Rhenium 186.21	76 **Os** Osmium 190.2	77 **Ir** Iridium 192.22	78 **Pt** Platinum 195.08	79 **Au** Gold 196.97	80 **Hg** Mercury 200.59	81 **Tl** Thallium 204.38	82 **Pb** Lead 207.2	83 **Bi** Bismuth 208.98	84 **Po** Polonium 208.98	85 **At** Astatine 209.99	86 **Rn** Radon 222.02
7	87 **Fr** Francium 223.02	88 **Ra** Radium 226.03	89-103 **Ac-Lr** #	104 **Rf** Rutherfordium 261.11	105 **Db** Dubnium 262.11	106 **Sg** Seaborgium 263.12	107 **Bh** Bohrium 262.12	108 **Hs** Hassium (265)	109 **Mt** Meitnerium (266)	110 **Uun** Ununnilium (269)	111 **Uuu** Unununium (272)	112 **Uub** Ununbium (277)		114 **Uuq** Ununquadium (289)		116 **Uuh** Ununhexium (289)		

★	57 **La** Lanthanum 138.91	58 **Ce** Cerium 140.12	59 **Pr** Praseodymium 140.91	60 **Nd** Neodymium 144.24	61 **Pm** Promethium 146.92	62 **Sm** Samarium 150.36	63 **Eu** Europium 151.96	64 **Gd** Gadolinium 157.25	65 **Tb** Terbium 158.93	66 **Dy** Dysprosium 162.50	67 **Ho** Holmium 164.93	68 **Er** Erbium 167.26	69 **Tm** Thulium 168.93	70 **Yb** Ytterbium 173.04	71 **Lu** Lutetium 174.97
#	89 **Ac** Actinium 227.03	90 **Th** Thorium 232.04	91 **Pa** Protactinium 231.04	92 **U** Uranium 238.03	93 **Np** Neptunium 237.05	94 **Pu** Plutonium 244.06	95 **Am** Americium 243.06	96 **Cm** Curium 247.07	97 **Bk** Berkelium 247.07	98 **Cf** Californium 251.08	99 **Es** Einsteinium 252.08	100 **Fm** Fermium 257.10	101 **Md** Mendelevium 258.10	102 **No** Nobelium 259.10	103 **Lr** Lawrencium 260.10

Key:

Atomic Number	3
Symbol	**Li**
Name	Lithium
Average Atomic Mass	6.94

Notes